T0189334

Intelligent Systems Reference Library

Volume 131

Series editors

About this Series

The aim of this series is to publish a Reference Library, including novel advances and developments in all aspects of Intelligent Systems in an easily accessible and well structured form. The series includes reference works, handbooks, compendia, textbooks, well-structured monographs, dictionaries, and encyclopedias. It contains well integrated knowledge and current information in the field of Intelligent Systems. The series covers the theory, applications, and design methods of Intelligent Systems. Virtually all disciplines such as engineering, computer science, avionics, business, e-commerce, environment, healthcare, physics and life science are included.

More information about this series at http://www.springer.com/series/8578

Charalampos Konstantopoulos
Grammati Pantziou
Editors

Modeling, Computing and Data Handling Methodologies for Maritime Transportation

Editors
Charalampos Konstantopoulos
Department of Informatics
University of Piraeus
Piraeus
Greece

Grammati Pantziou
Department of Informatics
Technological Educational Institute
of Athens
Athens
Greece

ISSN 1868-4394 ISSN 1868-4408 (electronic)
Intelligent Systems Reference Library
ISBN 978-3-319-87166-0 ISBN 978-3-319-61801-2 (eBook)
DOI 10.1007/978-3-319-61801-2

Softcover reprint of the hardcover 1st edition 2017

Printed on acid-free paper

This Springer imprint is published by Springer Nature
The registered company is Springer International Publishing AG
The registered company address is: Gewerbestrasse 11, 6330 Cham, Switzerland

The vast majority of human beings dislike and even actually dread all notions with which they are not familiar... Hence it comes about that at their first appearance innovators have generally been persecuted, and always derided as fools and madmen.

from "Words of Wisdom" by Aldous Huxley

To our families.

Foreword

Innovation is the keyword that drives and brings together the scientific and commercial world these days. Computer science and technologies are usually hidden beneath the success, especially in the disciplines well established before IT revolution in the end of the last century. This is exactly the case in maritime transportation: Ideas of e-navigation, GPS-based logistics, artificial intelligence in routing, or cost optimization are only mere examples of what can be achieved when traditional transportation approaches are boosted by aptly applied computer technologies. Despite the fact that a significant part of maritime industry has already been somehow IT-supported, there are still issues and aspects waiting to be solved, implemented, deployed, or improved.

A well-known problem of any interdisciplinary research is a lack of common understanding between the specialists. In technical sciences, often there are multiple discipline-specific terms that might be unclear to researches from other fields. An AI solution applied to marine technology may be misunderstood by IT researchers as well as mariners. The former might not be aware of the problem's roots and organizational or legal constraints (e.g., COLREGs), while the latter would be puzzled with terms such as "metaheuristic". Thus, it is crucial to establish a communication platform and knowledge exchange. Any interdisciplinary meeting or publication is a step forward. And this precisely is the root of the "Modeling, Computing and Data Handling Methodologies for Maritime Transportation" volume.

This volume comprises of seven chapters proposing solid IT-based models and systems in various maritime transportation fields, ranging from shipping safety and security issues to maritime logistics optimization. Three of the chapters are extended versions of papers from the 1st Workshop on Modeling, Computing and Data Handling for Marine Transportation (MCDMT 2015) in association with the 6th International Conference on Information, Intelligence, Systems and Applications (IISA 2015), July 6–8, 2015, Corfu, Greece. Such workshops as

MCDMT are great opportunities to fully integrate transportation and computer science worlds and may contribute to innovation breakthrough in the field. I am happy to notice such initiatives and waiting in anticipation what the next edition of the workshop will bring.

Joanna Szłapczyńska
Division of Computer Science Foundations and Computer Networks
Department of Navigation
Faculty of Navigation
Gdynia Maritime University
Gdynia, Poland

Preface

Maritime transportation is a major conduit of international trade. In terms of cost, maritime transport is very competitive against land and airborne transport, increasing only by a few percent the total product cost. On the other hand, it takes longer and may cause harbor congestion which may further increase the voyage time. Furthermore, there are difficulties in integrating this transportation mode efficiently with other transport or distribution options. On top of these, the safety and the environmental impact of maritime transportation, in particular in the case of sea accidents, are always two challenging issues.

Recent advances on maritime transportation require the synergy of both computer and maritime science. Computational intelligence, data mining and knowledge discovery/representation, risk assessment methodologies, as well as combinatorial optimization are the IT fields that have gained significant interest in maritime studies because of their potential in giving solutions for effective sea transportation. This edited volume focuses on research works related to the latest developments of IT methodologies for maritime transportation, and it comes after the successful organization of the the 1st and 2nd Workshop on Modeling, Computing and Data Handling for Marine Transportation (MCDMT 2015 and MCDMT 2016) which were held in association with the 6th and 7th International Conference on Information, Intelligence, Systems and Applications (IISA 2015 and IISA 2016). Seven chapters describing modeling tools, methodologies, algorithms, and systems comprise this edited volume as follows:

Chapters 1 and 2 consider two important problems in maritime logistics pertaining to quayside operational planning. Quayside problems include the Berth Allocation Problem (BAP) which determines the berths that incoming vessels are assigned to, the Quay Crane Assignment Problem, whereby the required cranes are assigned to each ship, and the Quay Crane Scheduling Problem (QCSP) where scheduling of crane tasks takes place. Chapter 1 considers a variation of BAP, namely the Minimum Cost Hybrid BAP (MCHBAP) with fixed handling times of vessels. The objective function to be minimized includes the cost of positioning, the speeding up or waiting, and the tardiness of completion for all vessels. A number of metaheuristics are surveyed, and a general variable neighborhood search approach is proposed. The metaheuristics are evaluated on real-life and randomly generated instances. Chapter 2 considers the problem of vessel stability during the process of unloading and/or loading containers onto vessels. The quay crane scheduling process determines the operational profile of each quay crane in terms of the container tasks and timing. The literature on the QCSP and related problems pertaining to quayside operational planning is surveyed considering vessel stability constraints to allow for quay crane schedules that can be used in practice, and directions are provided for future work in the area.

Chapters 3 and 4 focus on maritime routing problems. Specifically, Chap. 3 presents an extensive computational study of simple, but prominent metaheuristics to find high-quality ship schedules and inventory policies for a class of maritime inventory routing problems. Several variants of rolling horizon heuristics, K-opt heuristics, local branching, solution polishing, and hybrid metaheuristics are compared. Many of them substantially outperform the commercial mixed-integer programming solvers. Chapter 4 presents evolutionary algorithms for solving the real-time ship weather routing problem. The objectives to be minimized are the mean total risk and the fuel cost incurred along the obtained route while considering the time-varying sea and weather conditions and also a constraint on the total voyage time. The proposed approaches return only solutions compliant with the guidelines of the International Maritime Organization (IMO) and are tested on real data and also compared with an exact algorithm which solves the same problem.

Chapters 5 and 6 present decision support systems for safe shipping and seaport's security. In particular, Chap. 5 describes a decision support tool for environmentally safe shipping focusing on extracting aggregated statistics using spatial analysis of multilayer information, namely vessel trajectories, vessel data, and information regarding environmentally important areas. The proposed system includes preprocessing, clustering of trajectories based on their spatial similarity, and risk assessment employing probabilistic models. Applications are presented in areas such as queries in protected areas and marine traffic monitoring for environmental safety. Chapter 6 presents a decision support system for the assessment of seaports' security employing a flexible approach to evaluate the performance of security measures. A fuzzy analytical hierarchy process is utilized to analyze the complex structure of a seaport system and determine the weights of security measures while evidential reasoning is used to synthesize the risk analysis. The approach may provide analysts with a flexible tool to develop and employ robust resilience strategies aimed at enhancing seaport security in a systematic manner.

Finally, Chap. 7 presents a step-by-step development of a model which simulates maritime traffic in Bosporus, Turkey. The model demonstrates the relationships between sea traffic rules, number of pilots, and waiting times. It is expected that the presentation of the process for building a simulation model will be a useful guide for model builders in the maritime transportation domain.

From our position, we wish to thank Prof. George Tsichritzis for his constant support and help during the preparation of this volume. We would also like to thank all the authors for their contributions and the reviewers for their assessment of the chapters. We hope that the readers will find the contents of this edited volume interesting and useful.

Piraeus, Greece
Athens, Greece

Charalampos Konstantopoulos
Grammati Pantziou

Contents

Contributors

Noura Al-Dhaheri Maqta Gateway, Abu Dhabi Ports, Abu Dhabi, United Arab Emirates

Eleni Charou Institute of Informatics and Telecommunications, National Centre for Scientific Research "DEMOKRITOS", Aghia Paraskevi, Greece

Myun-Seok Cheon Corporate Strategic Research, ExxonMobil Research and Engineering Company, Annandale, NJ, USA

Tatjana Davidović Mathematical Institute, Serbian Academy of Sciences and Arts, Belgrade, Serbia

Theodore Giannakopoulos Institute of Informatics and Telecommunications, National Centre for Scientific Research "DEMOKRITOS", Aghia Paraskevi, Greece

Murat M. Gunal Industrial Engineering Department, Turkish Naval Academy, Tuzla, Istanbul, Turkey

Sotirios Gyftakis Institute of Informatics and Telecommunications, National Centre for Scientific Research "DEMOKRITOS", Aghia Paraskevi, Greece

Stuart Harwood Corporate Strategic Research, ExxonMobil Research and Engineering Company, Annandale, NJ, USA

Andrew John Liverpool Logistics, Offshore and Marine Research Institute, Liverpool John Moores University, Liverpool, UK

Charalampos Konstantopoulos Department of Informatics, University of Piraeus, Piraeus, Greece

Ioanna Koromila Institute of Nuclear and Radiological Sciences and Technology, Energy and Safety, National Centre for Scientific Research "DEMOKRITOS", Aghia Paraskevi, Greece; Ship Dynamics, Stability and Safety Research Group, Department of Naval Architecture and Marine Engineering, National Technical University of Athens, Athens, Greece

Nataša Kovač Maritime Faculty, University of Montenegro, Kotor, Montenegro

George L. Nemhauser H. Milton Stewart School of Industrial and Systems Engineering, Georgia Institute of Technology, Atlanta, GA, USA

Zoe Nivolianitou Institute of Nuclear and Radiological Sciences and Technology, Energy and Safety, National Centre for Scientific Research "DEMOKRITOS", Aghia Paraskevi, Greece

Grammati Pantziou Department of Informatics, Technological Educational Institution of Athens, Athens, Greece

Dimitri J. Papageorgiou Corporate Strategic Research, ExxonMobil Research and Engineering Company, Annandale, NJ, USA

Stavros Perantonis Institute of Informatics and Telecommunications, National Centre for Scientific Research "DEMOKRITOS", Aghia Paraskevi, Greece

Ramin Riahi Liverpool Logistics, Offshore and Marine Research Institute, Liverpool John Moores University, Liverpool, UK

Zorica Stanimirović Faculty of Mathematics, University of Belgrade, Belgrade, Serbia

Francisco Trespalacios Corporate Strategic Research, ExxonMobil Research and Engineering Company, Annandale, NJ, USA

Aphrodite Veneti Department of Informatics, University of Piraeus, Piraeus, Greece

Jin Wang Liverpool Logistics, Offshore and Marine Research Institute, Liverpool John Moores University, Liverpool, UK

Zaili Yang Liverpool Logistics, Offshore and Marine Research Institute, Liverpool John Moores University, Liverpool, UK

Abbreviations

ADP	Approximate Dynamic Programming
AHP	Analytical Hierarchy Process
AIS	Automatic Identification System
B&P	Branch and Price
BAP	Berth Allocation Problem
BCO	Bee Colony Optimization
BPGS	Branch-and-Price Guided Search
CBP	Customs and Border Protection agency
CDDA	Common Database on Designated Area
CMI	Critical Maritime Infrastructure
COTS	Commercials Off the Shelf
CRP	Container Relocation Problem
CSI	Container Security Initiative
CSO	Company Security Officer
CTOS	Container Terminal Operating System
C-TPAT	Customs-Trade Partnership Against Terrorism Initiative
DBAP	Discrete Case Berth Allocation Problem
DES	Discrete Event Simulation
DNV	Det Norske Veritas
DTED	Digital Terrain Elevation Data
EA	Evolutionary Algorithm
EDI	Electronic Data Interchange
ER	Evidential Reasoning
ES	Event Scheduling
ETA	Event Tree Analysis
FAHP	Fuzzy Analytical Hierarchy Process
FSA	Formal Safety Assessment
FTA	Fault Tree Analysis
GA	Genetic Algorithm
GIS	Geographical Information System

GVNS	General Variable Neighborhood Search
HBAP	Hybrid layout Berth Allocation Problem
IAT	Inter-Arrival Time
IED	Improvised Explosive Device
IMO	International Maritime Organization
IMTS	Index of Maritime Traffic Situation
IPT	Inter-Ping Time
IRP	Inventory Routing Problem
ISF	Importer Security Filing
ISPS	International Shipboard and Port Facility Security
KML	Keyhole Markup Language
LNG	Liquefied Natural Gas
LPG	Liquefied Petroleum Gas
LRIT	Long-Range Identification and Tracking System
MADA	Multiple Attribute Decision Analysis
MaritimeSim	Maritime Simulation Model
MCHBAP	Minimum Cost Hybrid Berth Allocation Problem
MILP	Mixed-Integer Linear Programming
MIP	Mixed-Integer Program
MIRP	Maritime Inventory Routing Problem
MIRPLib	Maritime Inventory Routing Problem Library
MOGA	Multi-Objective Genetic Algorithm
MSRAM	Maritime Security Risk Analysis Model
MT	Maritime Transportation
MTSA	Maritime Transportation Security Act
MTSS	Maritime Traffic Simulation System
NII	Non-Intrusive Inspection
NSGA	Non-dominated Sorting Genetic Algorithm
OR/MS	Operational Research/Management Science
OVCF	Ocean Vessel Carrier Filing
PFSO	Port Facilities Security Officer
PRA	Probabilistic Risk Analysis
QC	Quay Crane
QCAP	Quay Crane Assignment Problem
QCASP	Quay Crane Assignment and Scheduling Problem
QCSP	Quay Crane Scheduling Problem
RFID	Radio Frequency Identification
RHH	Rolling Horizon Heuristic
SC	Straddle Carrier
SPEA	Strength Pareto Evolutionary Algorithm
TFN	Triangular Fuzzy Number
TSS	Traffic Separation Scheme
TSVTS	Turkish Strait Vessel Traffic Service
VGO	Vacuum Gas Oil
VMI	Vendor-Managed Inventory

VND	Variable Neighborhood Descent
VTIS	Vessel Tracking and Information System
VTMIS	Vessel Traffic Management and Information System
WMD	Weapons of Mass Destruction

Chapter 1
Metaheuristic Approaches for the Minimum Cost Hybrid Berth Allocation Problem

Nataša Kovač, Zorica Stanimirović and Tatjana Davidović

Abstract The Minimum Cost Hybrid Berth Allocation problem is defined as follows: for a given list of vessels with fixed handling times, the appropriate intervals in berth and time coordinates have to be determined in such a way that the total cost is minimized. The costs are influenced by positioning of vessels, time of berthing, and time of completion for all vessels. Having in mind that the speed of finding high-quality solutions is of crucial importance for designing an efficient and reliable decision support system in container terminal, metaheuristic methods are the obvious choice for solving MCHBAP. In this chapter, we survey Evolutionary Algorithm (EA), Bee Colony Optimization (BCO), and Variable Neighborhood Descent (VND) metaheuristics, and propose General Variable Neighborhood Search (GVNS) approach for MCHBAP. All four metaheuristics are evaluated and compared against each other and against exact solver on real-life and randomly generated instances from the literature. The analysis of the obtained results shows that on instances reflecting real-life situations, all four metaheuristics were able to find optimal solutions in short execution times. The newly proposed GVNS showed to be superior over the remaining three metaheuristics in the sense of running times. Randomly generated instances were out of reach for exact solver, while EA, BCO, VND, and GVNS easily provided high-quality solutions in each run. The results obtained on generated data set show that the newly proposed GVNS outperformed EA, BCO, and VND regarding the running times while preserving the high quality of solutions. The computational analysis indicates that MCHBAP can be successfully addressed by GVNS and we believe that it is applicable to related problems in maritime transportation.

N. Kovač
Maritime Faculty, University of Montenegro, Dobrota 36, 85330 Kotor, Montenegro
e-mail: knatasa@ac.me

Z. Stanimirović (✉)
Faculty of Mathematics, University of Belgrade, Studentski trg 16,
11000 Belgrade, Serbia
e-mail: zoricast@matf.bg.ac.rs

T. Davidović
Mathematical Institute, Serbian Academy of Sciences and Arts, Kneza Mihaila 36/III,
11000 Belgrade, Serbia
e-mail: tanjad@mi.sanu.ac.rs

C. Konstantopoulos and G. Pantziou (eds.), *Modeling, Computing and Data Handling Methodologies for Maritime Transportation*, Intelligent Systems Reference Library 131, DOI 10.1007/978-3-319-61801-2_1

Keywords Container terminal · Scheduling vessels · Penalties · Evolutionary algorithm · Bee-colony optimization · Variable neighborhood search

1.1 Introduction

The effective and economic usage of port resources represents one of the most important issues in maritime transportation. Among all studied topics in the optimization of maritime transportation, Berth Allocation Problem (BAP) and its numerous variants take significant place. BAP assumes that the vessels have to be processed during a considered planning horizon while also taking into account the layout of the port. Vessels are essentially a set of following data: size, expected arrival time, estimated handling time, the most favored berth in the port, as well as the penalties that are used in the objective function. BAP has a goal to set each vessel in appropriate berth and time intervals in such a way that the cost function is optimized. NP-hardness of BAP is discussed in [25].

BAP has been intensively studied in recent literature [13, 20, 42]. Sometimes it is combined with another resources, called cranes, which are usually treated in the second stage of problem solving [33]. Various objective functions were considered for optimization. Some studies deal with minimization of the total waiting and handling times [1, 19], whereas the others consider the minimization of total costs for waiting and handling, and the costs of earliness or delay in completion as the objective [16]. The classification scheme for BAPs was proposed in [2] and elaborated in [3]. It is based on four attributes: spatial, temporal, handling time, and performance measure.

According to the spatial attribute, BAPs can be discrete, continuous, hybrid or draft. In the *discrete* case (DBAP), a quay is divided into a finite number of sections, denoted as berths. DBAP assumes that each berth can serve one vessel at a time and each vessel is allowed to occupy only one berth at a time. In addition, a given time horizon is also measured in discrete units. In the *continuous* case, a calling vessel can be allocated at any position in a quay with a requirement to avoid overlapping with other vessels. *Hybrid* layout (HBAP) is obtained in the BAP formulation if quay is partitioned in berths and if vessels can share one berth or vessels can occupy more than one berth. Finally, in the fourth layout, berthing position of a vessel also depends on its draft.

The most common BAP models with respect to the temporal attribute are *static* and *dynamic*. In the case of static BAP, the arrival times are either not specified or they may be changed during the berthing process, i.e., they are considered as soft constraints of the problem. The fact that the arrival times are not specified means that the considered vessels are already at the port and that they can be berthed instantaneously. The second case of static BAP considers the possibility that the vessels can be either slowed down or speeded up at a certain cost. In the dynamic BAP, the arrival times of the vessels are fixed, meaning that vessels cannot be berthed before their expected time of arrival.

According to the handling time attribute, BAPs are classified in four categories: BAPs with fixed handling times, with handling times depending on the berthing position, on the assignment of Quay Cranes (QCs), or on a QC operation schedule. The last attribute actually corresponds to the objective function of a considered BAP. The value of the objective function can depend on waiting time of a vessel, handling time of a vessel, completion time of a vessel, speedup of a vessel to reach the terminal before the expected arrival time, tardiness of a vessel against the given due date, berthing of a vessel apart from its desired berthing position, and some other factors. Detailed survey of BAP variants can be found in [3, 36].

Various solution approaches to BAP have been proposed in the literature up to now. In practice, it is important to have an efficient decision support system for helping the port manager to trade-off between the quickness of vessel service and the economic usage of the vessels at the port. When it is taken into the consideration that both container vessels and the port's resources impose great costs, it is beneficial for them to be utilized in the best possible way. Ports are highly dynamic systems and the manager usually has to make the decision in very short time. Situation in the port is sometimes changing on a minute basis, and therefore, seconds can be crucial for making a right decision. Because of this, it is of high importance to develop efficient optimization algorithm that will provide the manager with necessary data. Therefore, only few papers propose exact methods as solution approach [32], mostly in some special considerations of BAP. On the other hand, the majority of papers use heuristic methods, such as Lagrangean relaxation [19], Branch-and-Bound-based heuristic [1], etc. Having in mind the complexity of the problem and the need for efficient providing a high-quality solution, the metaheuristic methods, such as simulated annealing [29], tabu search [6], ant colony optimization [5], particle swarm optimization [41], etc., are the natural choice for BAP and its variants. Detailed overview of the metaheuristic approaches for BAP can be found in [23].

In this chapter, the static Minimum Cost Hybrid Berth Allocation Problem (MCHBAP) is considered. This variant of BAP does not involve QC scheduling, as it is assumed that each berth is equipped by a crane. According to the notation from [2], MCHBAP is classified as $hybr|stat|fix|\sum(w_1\,pos + w_2\,speed + w_3\,wait + w_4\,tard)$. The value for spatial attribute is set to *hybr*, as the model considered in this work is inspired by the model of [35] and it corresponds to the case shown in Fig. 3d from [2]. The objective function is a weighted sum of four components: berthing of a vessel apart from its favored berthing position, speeding up or waiting of a vessel with respect to the expected arrival time, and tardiness of a vessel against its due dates. It is adapted from paper [35] dealing with static MCHBAP.

Regarding solution approaches to MCHBAP, the Mixed Integer Linear Programming (MILP) formulation is proposed in [8] and it was used within the CPLEX commercial software and three MIP-based heuristics. However, due to the formulation complexity, none of the MIP based heuristics was able to give satisfactory results for problem instances involving more than 20 vessels. These results indicate that a metaheuristic would be a promising approach for larger problem instances of MCHBAP. In this chapter, we provide a survey of three metaheuristic approaches for MCHBAP: Bee Colony Optimization (BCO) [22], Evolutionary algorithm (EA)

[24], and Variable Neighborhood Descent (VND) [7] and propose General Variable Neighborhood Search (GVNS) as a new one.

Evolutionary algorithms are adaptive, global search techniques that exploit the concepts of natural adaptation and selective breeding of organisms [17]. EAs work with a population of individuals, where each individual represents a possible solution of the given problem. The main goal of EA is to increase the quality of individuals and the whole population by iteratively applying genetic operators, such as selection, crossover, mutation, etc. In the literature, there are numerous studies that propose EAs as solution methods to different variants of BAP. Goh and Lim [12] considered a NP-complete variant of BAP, named as Ship Berthing Problem. The performance of three metaheuristic methods designed for this variant of BAP was investigated in [12]: Randomized Local Search, Tabu Search and Genetic Algorithms. The problem of determining a dynamic berth assignment in the public berth system was considered by Nishimura et al. [31]. A Genetic Algorithm was proposed as solution method, producing good quality solutions in reasonable amount of computational time. Imai et al. [18] introduced the variant of BAP at a multi-user container terminal with indented berths, formulated it as a integer linear program, and designed a GA method to solve the considered problem. The discrete dynamic berth scheduling problem that minimizes the total service time of all the vessels was studied by Han et al. [14]. A hybrid method, named as GASA, representing a combination of Genetic Algorithm and Simulated Annealing, was proposed as solution approach in [14]. The computational results showed that GASA outperformed pure GA in the sense of solution quality and running times. In [4], a hybrid parallel GA, which combined parallel GA and heuristic algorithm, is applied to a dynamic berth allocation and quay crane assignments. An GA-based optimization heuristic was first proposed in the literature by Theofanis et al. [40] when solving the dynamic berth scheduling problem. Another EA-based optimization method was designed by [24] to solve MCHBAP with fixed handling times of vessels. As in many variants of BAP, the main problem when implementing a GA method for MCHBAP is large number of infeasible solutions that may appear during the GA run, mostly when applying crossover operator. For this reason, the EA-based method proposed in [24] excludes crossover as variation operators. In order to preserve the diversity of individuals in the population and to increase the quality of solutions, the authors of [24] involved four types of mutation operator and a two-phase improvement strategy in the designed EA-based approach.

BCO is an optimization algorithm based on the behavior of honey bees in nature. The core idea of BCO is to explore the principles which bees use when searching for food in nature, and to design optimization algorithms that would efficiently search for the optimal solution of the given combinatorial optimization problem [9]. In the literature, one can find numerous examples of successful applications of BCO to various transportation, location, and scheduling problems, majority of them being summarized in [39]. In [22], the constructive variant of BCO for MCHBAP is developed and it is proven that the optimal solution is reachable. To enhance the performance of the BCO algorithm, three improvement techniques are proposed and their complexities are discussed.

VNS is a metaheuristic method based on local search procedure, introduced by Mladenović and Hansen [28]. The main components of VNS are *shaking* and *local search*, which are iteratively applied together with a systematic change of neighborhoods [15]. The role of shaking phase is to help the algorithm to escape from the local minimum and to explore the search space in an efficient manner. Local search is used to intensify exploration of neighborhoods around promising solutions. VNS showed to be successful metaheuristic method for various combinatorial and global optimization problems [15]. Hansen et al. [16] proposed VNS implementation adapted to minimum cost discrete BAP. The variant of BAP considered in [16] penalizes any late departures of the vessels and awards all early departures. As the discrete BAP studied in [16] assumes that each vessel occupies only one berth, it is simpler than MCHBAP. The authors define several neighborhood structures for VNS in accordance with the characteristics of the considered discrete BAP. The results of computational study conducted in [16] show that VNS is able to provide optimal solutions for all small size instances and to outperform existing algorithms on large instances. The study [7] considers MCHBAP and proposes Variable Neighborhood Descent (VND) as a new optimization approach to this problem, VND actually represents a deterministic variant of Variable Neighborhood Search method. VND does not involve shaking component, while the local search is performed through multiple neighborhoods. Three types of neighborhoods are defined, based on *sequence pair* solution representation, and used within VND method. In this chapter, we propose a General Variable Neighborhood Search (GVNS) for MCHBAP, which is developed in order to ensure better diversification of solutions compared to VND. The proposed GVNS employs shaking procedure based on stochastic transformations of the best solution. Instead of simple local search, GVNS employs VND to improve the solution by systematic exploration of six neighborhoods of the current solution. The superiority of the proposed GVNS over VND from [7], EA from [24], and BCO from [22] developed for MCHBAP is confirmed through the set of computational experiments.

The remaining part of this chapter is structured as follows. The description of the considered MCHBAP is provided in Sect. 1.2. In Sects. 1.3, 1.4 and 1.5, we provide detailed description of EA, BCO and two VNS-based metaheuristic for MCHBAP, respectively. Experimental results and analysis are presented in Sect. 1.6. Final remarks and suggestions for future research are contained in Sect. 1.7.

1.2 Minimum Cost Hybrid Berth Allocation Problem

For a given set of incoming vessels that aim to be served within a pre-specified planning horizon, a solution of MCHBAP defines a berthing position and a berthing time in such a way that the total berthing cost is minimized. This cost for each vessel consists of four components: costs of positioning a vessel apart from its favored berth, speeding up or waiting caused by missing the estimated arrival time of a vessel, and tardiness of completion for a vessel. It is assumed that the handling (operation)

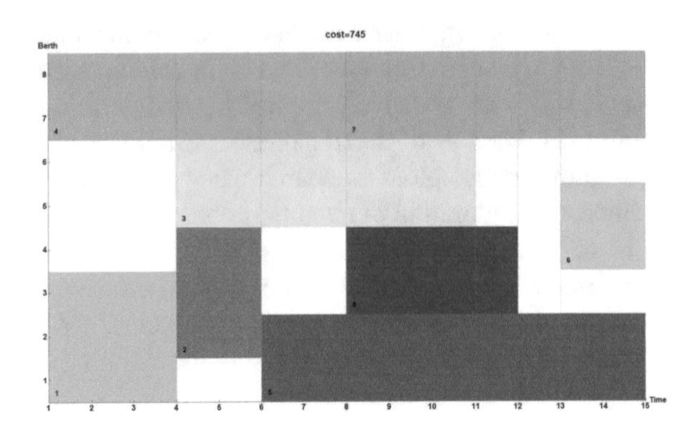

Fig. 1.1 An illustration of BAP solution

times of all vessels are known input parameters. As illustrated in Fig. 1.1, a planning horizon can be presented by a rectangle with a given space and time coordinates. A solution of MCHBAP defines packing small rectangles (representing vessels) into the given planning horizon. It is assumed that both coordinates are discrete, as space is modeled by the berth indices whereas the time is divided into segments in such a way that berthing time of each vessel can be represented by an integer. Furthermore, the height of each rectangle representing a vessel is equal to the length of the corresponding vessel (expressed by the number of berths), while its width corresponds to the handling time of that vessel. The *reference point* of a vessel (marked by the index of vessel in Fig. 1.1) is given in the lower-left corner of a rectangle and its coordinates are equal to berthing position and berthing time of the corresponding vessel. An allocation is *feasible* if all rectangles can fit in the given planning horizon without overlapping (Fig. 1.1).

1.2.1 Problem Description

MCHBAP is described by the input data, objective function, and a set of constrains that define feasible solutions. The input data of MCHBAP are listed below:

l Number of vessels;
m Number of berthing positions;
T Number of time segments in the planning horizon;
$vessel$ A set of 9-tuples containing data needed to describe all l vessels, i.e.,

$$vessel = \{vessel_k | k = 1, \ldots, l\}.$$

where

$$vessel_k = \{ETA_k, a_k, b_k, d_k, s_k, c_{1k}, c_{2k}, c_{3k}, c_{4k}\}, k = 1, \ldots, l.$$

The elements of $vessel_k$ are described as follows::

ETA_k	The expected arrival time of $vessel_k$;
a_k	Processing time of $vessel_k$ when one QC is used;
b_k	Length of $vessel_k$ expressed as the number of berths;
d_k	Due departure time for $vessel_k$;
s_k	Favored berthing location of the reference point of $vessel_k$;
c_{1k}	Penalty cost, if $vessel_k$ misses its favored berth;
c_{2k}	Penalty cost per unit time, if $vessel_k$ is speeded up with respect to ETA_k;
c_{3k}	Penalty cost per unit time, if $vessel_k$ waits or slows down with respect to ETA_k;
c_{4k}	Penalty cost per unit time, if $vessel_k$ cannot departure before its due time d_k.

A feasible solution of MCHBAP consists of pairs $(B_k, At_k), k = 1, 2, \ldots, l$, where B_k denotes the lowest berth index allocated to the $vessel_k$, $B_k \in \{1, 2, \ldots, m\}$ and At_k represents the minimum time index for the $vessel_k$, $At_k \in \{1, 2, \ldots, T\}$. Pair (B_k, At_k) actually corresponds to the reference point of $vessel_k$, $k = 1, 2, \ldots, l$. The feasible solution of MCHBAP has to satisfy the following sets of constraints:

Constraints 1 At each time segment t, $t = 1, \ldots, T$, a berth can be assigned to only one vessel,

Constraints 2 A berth can be allocated to a vessel only between its arrival and departure times.

The objective function that our MCHBAP aims to minimize is the total penalty cost including: the penalty of missing the favored berthing location of the reference point, the penalties of berthing earlier or later than the expected arrival time, and the penalty of departure after the promised due time. Precisely, mathematical expression of the objective function is as follows (see [35]):

$$\sum_{k=1}^{l} \left(c_{1k}\sigma_k + c_{2k}(ETA_k - At_k)^+ + c_{3k}(At_k - ETA_k)^+ + c_{4k}(Dt_k - d_k)^+ \right), \tag{1.1}$$

where

$$\sigma_k = \sum_{t=1}^{T} \sum_{i=1}^{m} \{|i - s_k| : \text{ position } (t, i) \text{ is assigned to vessel } k\},$$

$$(a - b)^+ = \begin{cases} a - b, & \text{if } a > b, \\ 0, & \text{otherwise,} \end{cases}$$

and $Dt_k = At_k + \lceil a_k/b_k \rceil$ represents the departure time of the $vessel_k$, $Dt_k \in \{1, 2, \ldots, T\}$. Namely, if only one crane is used to serve the $vessel_k$, the required processing time is a_k. However, if the $vessel_k$ occupies more berths (each equipped by a crane), the processing time will be reduced b_k times.

According to the definition of the objective function given in [35], σ_k is expressed by the given double sum. It can be explained as follows: as the $vessel_k$ can occupy several berths and only one is preferred (which usually means that it contains the required equipment for serving the vessel), all other allocated berths have to be penalized. The lack of the proper equipment on these berths requires the engagement of the additional equipment and/or labor. All that increases the costs of handling the vessel. Finally, $(a - b)^+$ denotes that term has impact only if its value is positive.

The MILP formulation of MCHBAP was proposed in [8] and used within CPLEX exact solver for solving small to medium size problem instances. In addition, it was incorporated into MIP-based heuristic methods to obtain sub-optimal solutions for the larger examples. From the MILP formulation proposed in [8], it may be concluded that the MCHBAP complexity regarding the number of variables is $O(mlT)$. It means that an exact solver must determine the values of $l \cdot T(2 \cdot m + 1)$ binary variables and $2 \cdot l$ integer ones in the optimal solution. In addition, it is necessary to calculate objective function value of the optimal solution. The MILP formulation from [8] also implies that the MCHBAP complexity in respect to the number of constraints is $O(lT^2 + mT + ml)$. Note that MCHBAP is NP-hard problem as a variant of BAP.

Experimental evaluation described in [8] shows that, due to the complexity of MILP formulation, the gaps between the obtained solutions and the optimal ones are large. Therefore, metaheuristic methods described in this chapter are based on combinatorial formulation described above.

1.2.2 Basic Definitions and Notations

Container terminal is represented as a two dimensional plane with time and berth as axes. Vessels are modeled by the rectangles of heights equal to the number of berths (b_k) and widths matching the needed handling time ($w_k = \lceil a_k/b_k \rceil$). Therefore, MCHBAP can be observed as a variant of rectangle packing problem.

All methods considered here are based on the combinatorial formulation for MCHBAP and use the same data structures and initialization (preprocessing) phase. During the initialization phase, the list of 3-tuples containing berth and time coordinates and the corresponding penalty cost as elements is created for each vessel. It is named ξ list and it covers all possible positions of the vessel in the planning horizon. The collection of all l individual ξ lists is denoted by Ξ.

The ξ lists are sorted in the non-decreasing order of the penalty cost values. The role of ξ lists is to assure efficient search through solution space by its significant reduction in each step. More precisely, any change in the allocation of vessels in port produces the changes in the corresponding ξ lists in such a way that only feasible positions for each vessel remain as the elements in its list. As ξ lists remain sorted at each step of the search, it is easy to detect positions with smaller penalty cost for each allocated vessel (if there is any). In addition, ξ list structure enables easy identification of unfeasible solution: if at any moment there exists a vessel with empty ξ list, the corresponding solution could be discarded as unfeasible.

During the execution of metaheuristic methods, the main steps are the selection of a vessel to be allocated and the selection of its position in berth-time plane.

These decisions are made stochastically, based on the priorities of vessels. Criterion for vessel selection in EA and BCO is a linear combination of the corresponding *ETA* parameter value, the size of the rectangle representing the corresponding vessel in the planning horizon, and the calculated average cost of all possible ξ list elements for the observed vessel. Coefficients of this linear combination are named λ_1, λ_2, and λ_3, respectively. Their values are determined experimentally in such a way that $\lambda_1 + \lambda_2 + \lambda_3 = 1$. The linear combination of parameters with the corresponding coefficients represents the priority of a vessel used for its selection by the roulette wheel.

The selection of vessels' positions in BCO and in the initialization phase of EA is performed stochastically, based on the position costs. The positions with smaller costs have higher chances to be selected.

All metaheuristic methods keep the best found solution, the corresponding total cost value, and the CPU time of first occurrence of the best found solution in global variables *Solution*, *GlobalBest*, and *minT*, respectively.

1.3 Evolutionary Algorithm for MCHBAP

In this section, we review Evolutionary algorithm (EA) for MCHBAP, introduced in [24]. Having in mind the problem characteristics, the EA from [24] uses integer encoding of individuals and it involves only four types of mutation operator as a variation part. The crossover operator is omitted from the EA for MCHBAP, as it often produces infeasible individuals. After mutation operators are performed, a two-phase optimization procedure is applied in order to improve the quality of few selected individuals in each generation. The first phase of the improvement procedure is based on changing the associated berth of a vessel, while the second one allows only perturbations of vessels order within the chosen berth. In the following subsections, all aspects of EA proposed in [24] are explained in detail.

1.3.1 Representation of Individuals

Integer representation of individuals is used in the EA proposed in [24], as this type of encoding showed to be the most appropriate to the considered MCHBAP. The code of each individual is a list that consists of m sublists. Each sublist refers to one berthing position and it consists of indices of the vessels whose reference points are allocated to this position. The elements of each sublist are sorted in order in which they are to be scheduled at the corresponding berthing position.

Let us consider the example taken from [24]. The input data of the problem with $m = 8$ berths, $l = 5$ vessels, and $T = 15$ time units are given in Table 1.1.

Table 1.1 Input data for the problem with $m = 8$ berths, $l = 5$ vessels and $T = 15$ time units

$vessel_k$	ETA_k	a_k	b_k	d_k	s_k	c_{1k}	c_{2k}	c_{3k}	c_{4k}
1	1	9	3	4	1	10	20	20	25
2	4	6	3	6	3	10	20	20	25
3	4	14	2	11	6	10	20	20	25
4	5	14	2	12	7	10	20	20	25
5	6	18	2	16	2	10	20	20	25

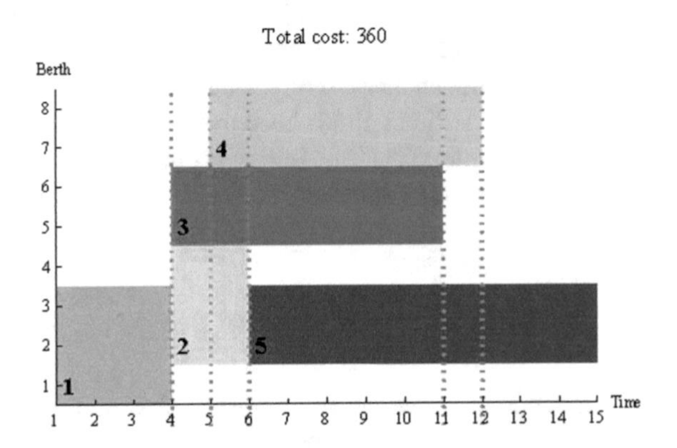

Fig. 1.2 Solution obtained by decoding *Individual_1*

The representation of an individual corresponding to a feasible solution to MCHBAP is as follows:

$$Individual\_1 : \{\underbrace{\{1\}}_{1}, \underbrace{\{5, 2\}}_{2}, \underbrace{\{\}}_{3}, \underbrace{\{\}}_{4}, \underbrace{\{3\}}_{5}, \underbrace{\{\}}_{6}, \underbrace{\{4\}}_{7}, \underbrace{\{\}}_{8}\}$$

The indices of vessels to be served on a berth are given in brackets, while the berth indices are indicated below the brackets. This representation means that berth 1 serves vessel 1, while vessels 5 and 2 are assigned to berth 2. Berths 5 and 7 serve vessels 3 and 4, respectively. Empty brackets above berth indices 3, 4, 6, and 8 imply that these berths are empty, i.e., no reference points of vessels are assigned to them.

Figure 2 from [24] illustrates the solution to MCHBAP obtained by decoding *Individual_1*. If the reference point of a vessel is assigned to the berth with index k, it means that this vessel belongs to the k-th sublist. As, the hybrid case of BAP is considered, it is allowed that a vessel occupies some space within more than one adjacent berths.

From Fig. 1.2, it can be seen that vessels 5 and 2 are served on berth 2, but the order on time axis does not follow the order of vessels in the individual's representation. Vessel 5 is allocated to port before vessel 2 and therefore, it is placed on the cheapest possible location. Vessel 2 is then allocated at the cheapest location that remained after vessel 5 is served. The representation of *Individual_1* implies that berth 3 is empty, however, in the decoded solution, berth 3 is not free, as it is occupied by some parts of vessels 1, 2 and 5. Finally, the decoded solution to MCHBAP consists of 3-tuples (berth, time, cost), each of them corresponding to the reference points of a vessel in the two dimensional plane. Note that *Individual_1* and the corresponding allocation presented in Fig. 1.2 correspond to the optimal solution for this example.

It should be emphasized that the representation of an individual may lead to the schedule that exceeds the berth and/or planning time horizon limits. If this is the case, the individual is considered as *infeasible*, an it is discarded from the population. As the length of vessels is known in advance (from the input data), it can be easily checked whether or not a berth can handle an unallocated vessel. On the other hand, feasible individuals can be uniquely decoded in the MCHBAP solution.

1.3.2 Generating Initial Population for EA

In [24], the initial population contains *nEA* individuals. The generation of initial individuals is performed by the procedure INITIALIZE with pseudo-code shown in Algorithm 1. The input parameters of this procedure are population size (*nEA*) and the size of tournament for selecting a berth that will be initially assigned to the considered vessel. For each of *nEA* individuals in the initial population, the procedure INITIALIZE forms the corresponding Ξ list.

Algorithm 1 Generating the initial EA population

```
procedure INITIALIZE(nEA, size)
// Generates initial population with nEA individuals
  i ← 1
  while i ≤ nEA do
    UnusedVessels ← {1, 2, ..., l}
    individual(i) ← m empty lists
    while UnusedVessels ≠ ∅ do
      ID ← ROULETTE(UnusedVessels, selectionCriteria)
      Berths ← Determine all subsets of subsequent berths
                 with enough free space for vessel ID
      berthID ← TOURNAMENTFORBERTHS(size, Berths)
      individual(i) ← Insert vessel ID on the last position on berthID
      UnusedVessels ← UnusedVessels\{ID}
    end while
    if FEASIBLE(individual(i)) then
      i ← i + 1
    end if
  end while
end procedure
```

The procedure INITIALIZE creates a new individual based on decision regarding the order of vessels selection and on which berth to place each vessel. Vessel selection is based on priorities defined as a weighted sum of three components with the coefficients values λ_i, $i = 1, 2, 3$, which are set to 0.12, 0.13, 0.75, respectively.

The priorities are determined based on the initial ξ lists (where all positions are considered feasible) and they do not change during the allocation phase.

The procedure ROULETTE considers all unused vessels and among them chooses one by one vessel by performing the roulette wheel selection. The probability that a vessel will be selected depends on its previously calculated priority. As the length of each vessel and the total number of time segments in the planing horizon are known in advance, it is easy to determine the subsets of subsequent berths with enough free space to handle the selected vessel. The procedure TOURNAMENTFORBERTHS is applied next, in order to decide which berth should be assigned to the selected vessel. The berth is selected among all possible berths with enough free space for the examined vessel by the means of tournament selection. After the vessel and its associate berth are chosen, the vessel is placed as the last one on the selected berth. More precisely, the index of a vessel is placed at the end of the corresponding sublist of the berth to which the vessel is assigned.

The described steps are repeated until all vessels are associated to the berths. After that, the individual is decoded by actual allocation of vessels to the positions in berth-time plane. If an unfeasible individual is created, it is removed from the population and a new one is formed until nEA feasible individuals are obtained. The feasibility of an individual is checked by procedure FEASIBLE.

1.3.3 EA Operators

The EA proposed in [24] uses elitist strategy in the generation replacement, meaning that certain number of individuals from one EA generation directly passes to the next one. These individuals are named *elite* ones and their role is to preserve high quality genetic material that will be used to generate new EA generation. In the EA implementation from [24], the number of elite individuals is set to $eliteNO = 1/3 * nEA$. The remaining $2/3 * nEA$ individuals that will take part in creating new EA generation are chosen by fine-grained tournament selection [11] with two tournament sizes. Tournaments of $size_1$ are performed $40\% * (2/3nEA)$ times, while remaining individuals are selected via tournaments of $size_2$, where $size_1 < size_2$ holds. The basic idea behind the applied selection operator is to increase the chances of lower quality individuals to take part in producing new generation. The implemented selection operator also prevents duplicated individuals from entering the next EA generation, and thus enabling the diversity of individuals in the EA population with an aim to keep EA away from local optimum trap.

Existing studies in the literature that use EA as solution approaches for various optimization problems involve different types of variation operators. For example, EAs from [26, 37] use crossover as variation operator, while [10, 34] apply only mutation. The study [24] proposes EA for MCHBAP that uses no crossover operator in the EA concept. The reason is in the fact that all examined types of crossover operator produced large number of infeasible individuals. There are two strategies that can be applied when an infeasible individual appears during the EA run, i.e., it can be either discarded or corrected to be feasible. However, as too many infeasible offspring appear after each of the examined crossover operator, discarding all infeasible individuals results in significant decrease of the population size. On the other hand, it turns that different strategies applied to repair infeasible offspring were time consuming, and therefore, significantly affected the overall efficiency of EA. In addition, the quality of the best EA solution was worse than in the case when no crossover is incorporated. For these reasons, the EA implementation for MCHBAP proposed in [24] involves only mutation operators, which are designed in accordance with problem characteristics. More precisely, four types of mutation operators are incorporated: *Insert*, *Inversion*, *Scramble* and *Swap*. All four operators are applied to each individual in the population with the given probability μEA, producing at most four offspring. Each of the implemented mutation operators enables changes in both berth and time coordinates of the selected vessels in the individual that is subject to mutation. Therefore, they can produce large changes in the considered individual and the significant change in the objective function value. The role of the applied mutation operators is to help in restoring the lost subregions of search space during the EA run and in preventing the algorithm to finish in a local optimum. The four types of mutation operators implemented in EA are as follows.

Insert mutation is illustrated in Fig. 1.3 [24]. This operator randomly selects two vessels in the individual that is subject to mutation and moves the second vessel behind the first one. In Fig. 1.3, vessels 3 and 2 are chosen vessel 2 is moved to follow the vessel 3 in the mutated individual.

Inversion mutation first selects two vessels at random and then inverts the part of the chromosome between them, including these two vessels. Figure 1.4 [24] shows the way in which inversion mutation works: vessels 2 and 1 are chosen and the whole substring between these two vessels is inverted. Vessels 2 and 1 are inverted as well.

Swap mutation randomly chooses two vessels from a chromosome and swaps their positions. This mutation type is illustrated in Fig. 1.5 [24]. Vessels 4 and 3 are selected and their positions are interchanged in the mutated individual.

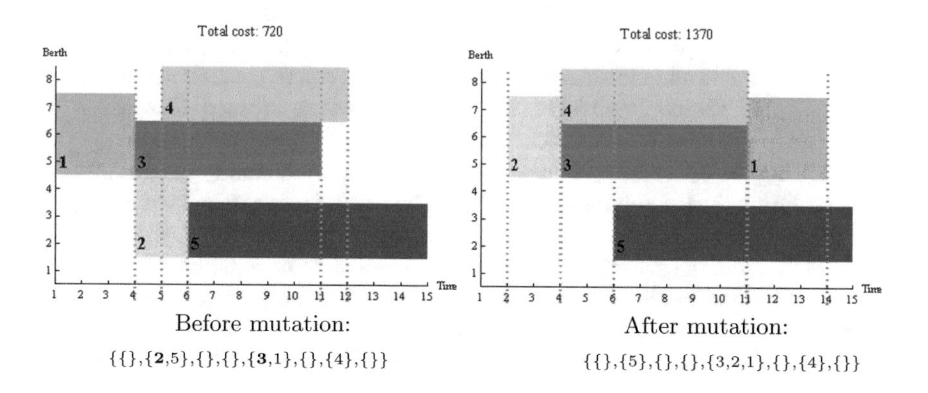

Fig. 1.3 An example of *Insert mutation*

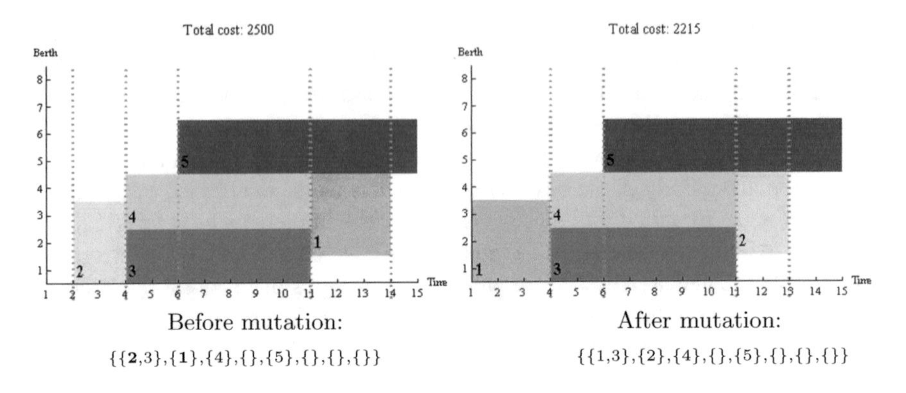

Fig. 1.4 An example of *Inversion mutation*

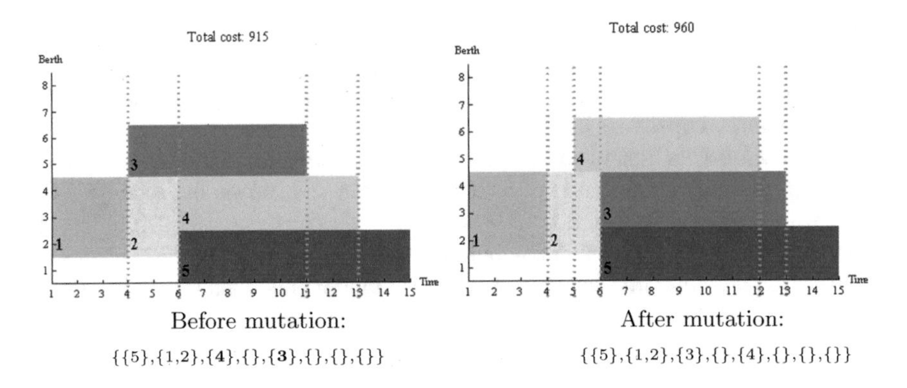

Fig. 1.5 An example of *Swap mutation*

Scramble mutation performs random permutation of the selected subset of vessels in the chromosome. The endpoints of the subset of vessels to be scrambled are randomly chosen from the individual's chromosome. They may belong to different sublists of vessels, and in some cases, it may happen that the whole chromosome is subject to scramble mutation. An example of applying scramble mutation is presented in Fig. 1.6 [24]. Vessels 1 and 5 are selected and all vessels between them, including the two endpoint vessels, are subject to random permutations.

Following the idea from [40], the value of mutation probability parameter μEA changes during the EA run for each of the applied mutation operators. At the beginning of the EA search, *Inverse* and *Scramble* mutations are favoured, while in later stages of EA run, the accent is put on *Insert* and *Swap* mutations. By applying this strategy, the algorithm first allows significant changes in the solution chromosomes. As the objective function value improves, the search is focussed on the solutions from a smaller region. The probability of applying *Scramble* and *Inversion* mutations on a gene is calculated as follows

$$\mu EA = 1 - 0.9 * (popID)/(maxGen), \tag{1.2}$$

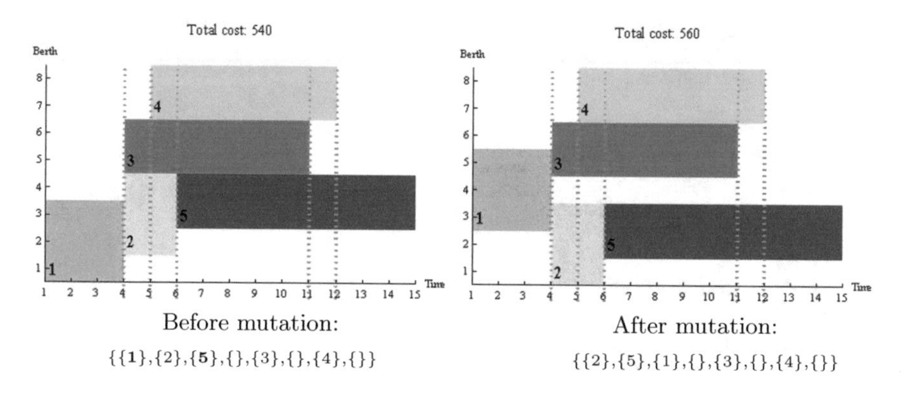

Fig. 1.6 An example of *Scramble mutation*

while the probability of *Insert* and *Swap* mutations is increasing according to the following formula

$$\mu EA = 0.9 * (popID)/(maxGen), \quad (1.3)$$

where *popID* denotes the index of the current EA generation, and *maxGen* represents the maximal number of EA generations. These formulae were determined through the set of preliminary experiments.

The EA from [24] involves an additional improvement strategy after the mutation part. The improvement procedure is applied on the subset of individuals in the population, which are selected by performing tournament selection. The number of individuals that are subject to improvement strategy, as well as the tournament size, are input parameters of EA. The level of selection pressure is tuned by varying the size of the tournament. Lager tournament size implies that lower fitted individuals will have smaller chances to be selected for improvement procedure, and vice versa.

The pseudo-code of the applied improvement procedure is given in Algorithm 2. The procedure IMPROVE tries to find better allocation of vessels on each berth in the selected individual. For each group, corresponding to a single berth, vessels are examined one by one, and they are allocated on the cheapest possible position on the given berth by procedure CHEAPESTALLOCATION. After all vessels are placed within corresponding berths, the procedure REORDER is applied to sort the vessels in the descending order in respect to their costs in the given allocation. The improvement procedure takes one by one vessel from the sorted array and examines its corresponding ξ list. If there are cheaper positions for a vessel, regardless the berth index, this vessel is moved to the cheapest available position. Cheaper positions for a vessel are detected by procedure EXISTSCHEAPER. At the end of the improvement process, the procedure MAKEINDIVIDUAL is called in order to make corresponding changes in the representation of the newly obtained individual. More precisely, the reallocated vessels are moved at the beginning of group corresponding to their new berths. The order of vessels within a group is changed according to performed reallocation, with an aim to first consider the vessels with the highest penalty costs.

Algorithm 2 Procedure for EA improvement phase

```
procedure IMPROVE(ind)
    for berth ← 1, m do
        group ← ind(berth)
        for j ← 1, LENGTH(group) do
            v ← group(j)
            sol(v) ← CHEAPESTALLOCATION(v, berth)
        end for
    end for
    vessels ← REORDER(sol)
    for i ← 1, l do
        if EXISTSCHEAPER(ξ(vessels(i)), sol(vessels(i))) then
            newSol(vessels(i)) ← ξ(vessels(i), 1)
        else
            newSol(vessels(i)) ← sol(vessels(i))
        end if
    end for
    newInd ← MAKEINDIVIDUAL(newSol)
    RETURN (newInd, newSol)
end procedure
```

In the second part of the EA improvement phase, a local search procedure is applied to each improved individual obtained as the result of procedure IMPROVE. Its pseudo-code is given in Algorithm 3. The applied local search algorithm reorders vessels on a single berth by calling procedure REARRANGE. The procedure CALCULATECOST determines the total cost of the corresponding group of vessels. If the total cost is reduced, the group reordering is accepted. The described steps are repeated for each group of vessels corresponding to a single berth. Note that in this step, new conflicts obtained by the perturbation of vessels are prevented. More precisely, vessels are allowed to change their positions only if the resulting allocation produces no conflict with vessels on adjacent berths.

Algorithm 3 Procedure for local search within EA

```
procedure BERTHLOCALSEARCH(ind, sol)
    for berth ← 1, m do
        group ← ind(berth)
        groupSol ← REARRANGE(group)
        if CALCULATECOST(groupSol) < CALCULATECOST(sol(group)) then
            sol(group) ← groupSol
        end if
    end for
    newInd ← MAKEINDIVIDUAL(sol)
    RETURN (newInd)
end procedure
```

After the EA improvement phase is completed, fitness values of individuals are calculated and the next generation is created based on the individuals' fitness values.

The above described steps are repeated until one of the EA termination criteria is satisfied. The EA uses the combination of two stopping criteria: time limit and maximal number of generations. On each improvement of the current best solution, the corresponding CPU time is saved in $minT$. The pseudo-code of EA for MCHBAP is presented in Algorithm 4, which is taken from [24].

Algorithm 4 Optimization Based EA for MCHBAP

```
procedure EAForMCHBAP
  Initialize(nEA)
  popID ← 1
  GlobalBest ← ∞
  while (SessionTime ≤ RunTime) ∧ (popID ≤ maxGen) do
    for each individual do
      μEA ← 0.9 * popID/maxGen
      newindividual1 ← InsertMutation(noGenes, μEA)
      newindividual2 ← SwapMutation(noGenes, μEA)
      newindividual3 ← InversionMutation(noGenes, 1 − μEA)
      newindividual4 ← ScrambleMutation(noGenes, 1 − μEA)
    end for
    for noImprovements individuals do
      individualID ← TournamentForImprovement(size_1, size_2, cost)
      {newindividualI, newSol} ← Improve(IndividualID)
      newindividualI ← BerthLocalSearch(newIndividualI, newSol)
    end for
    Calculate Cost for each individual
    if BestCost(popID) < GlobalBest then
      BestSolUpdate()
    end if
    eliteNO ← nEA/3
    Copy best eliteNO individuals to the new generation
    size1NO ← Round((nEA − eliteNO) * 0.4)
    Choose size1NO individuals through tournaments of size_1
    size2NO ← nEA − eliteNO − size1NO
    Choose size2NO individuals through tournaments of size_2
    popID ← popID + 1
  end while
end procedure
```

As it can be seen from the description of the EA for MCHBAP, the algorithm involves a number of parameters, such as population size, tournament sizes, stopping criteria parameters, etc. Therefore, a set of parameter tuning tests was performed to determine the values that ensure the best EA performance. Based on the obtained results, the values of EA parameters are chosen as follows. The number of individuals in population nEA is equal to 20, Tournament sizes $size_1$ and $size_1$ of the applied fine-grained tournament selection are set to 3 and 5, respectively. Formulae (1.2) and (1.3) describe how the probability of mutation μEA changes as the EA progresses. The number of individuals that are subject to the EA improvement phase is set to 5 in each generation. The maximal number of generations $maxGen$ is equal to 40, while the maximal running time *RunTime* is set to 10 min.

1.4 Bee Colony Optimization Algorithm for MCHBAP

The implementation of constructive BCO for MCHBAP proposed in [22] is recalled in this section. Population of B artificial bees is used to generate solutions of MCH-BAP through NC steps. Each step consists of *forward* and *backward* passes [9]. The remaining part of the section contains a detailed explanation of the implementation of BCO for MCHBAP from [22] including the three improvement techniques used to enhance the complete BCO solutions.

1.4.1 Implementation Details

The steps of a single iteration of BCO for MCHBAP are presented in Algorithm 5. At the beginning of the algorithm, a Ξ list and an initial (empty) solution are assigned to each bee. When generating partial solution, at each forward pass, the bee must make two decisions: to choose $\lceil l/NC \rceil$ vessels and to pick an allowed position from the ξ list for each of the chosen vessels.

Algorithm 5 An iteration of BCO

procedure ITERATION(B, NC)
  for $i \leftarrow 1, B$ **do**
    $Bee\Xi(i) \leftarrow start\Xi$
    $BeeSol(i) \leftarrow \{\}$
    $BeeCost(i) \leftarrow \infty$
  end for
  for $u \leftarrow 1, NC$ **do**
    for $b \leftarrow 1, B$ **do**
      for $j \leftarrow 1, \lceil l/NC \rceil$ **do**
        $v \leftarrow$ SELECTV($vessels$)
        $pos \leftarrow$ SELECTP($Bee\Xi(b)$)
        $BeeSol(b, v) \leftarrow pos$
        $Bee\Xi(b) \leftarrow$ UPDATELIST($Bee\Xi(b)$)
      end for
    end for
    for $b \leftarrow 1, B$ **do**
      $BeeCost(b) \leftarrow$ CALCULATECOST($BeeSol(b)$)
    end for
    RECRUITINGPROCESS(B)
  end for
end procedure

Vessels are selected in respect to their priorities with the coefficients $\lambda_i, i = 1, 2, 3$, taking values 0.8, 0, and 0.2, respectively. These values were determined by the preliminary experiments conducted in [22].

The method of selection of a new vessel to be added to the partial solution is based on the vessel's priority. It is implemented in SELECTV procedure, which uses random number generator and roulette wheel. For a selected vessel, all potential positions are considered with respect to the penalty cost values. Positions that would lead to a lower penalty costs are more likely to be selected. Each bee computes the probability for each position from the ξ list and picks one based on the randomly generated number and the roulette wheel by procedure SELECTP. The considered vessel is fixed on that position and procedure UPDATELIST is called to clear ξ lists for all unused vessels. The above described process is repeated $\lceil l/NC \rceil$ times by each bee, and then the forward pass is completed. The values of BCO parameters are determined experimentally in [22] and it was decided that both NC and B parameters should be set to 4. The $NC = 4$ actually means that up to $\lceil l/4 \rceil$ vessels are added to the partial solution in each forward pass.

In the initial stage of the backward pass [9], the procedure CALCULATECOST evaluates solutions of each bee, i.e., calculates the corresponding current cost. Based on the current cost, bee makes a decision on whether to keep the current arrangement of the vessels in the port and to recruit the other bees, or to become the uncommitted follower by abandoning its solution. If the bee fails to form the partial solution, i.e., if the final selected vessel can not be placed on a feasible position in its ξ list, the bee certainly becomes the uncommitted follower. The cost of unfeasible solution is set to infinity and the corresponding bee becomes an uncommitted follower. Smaller current cost corresponds to larger probability of the bee staying loyal to its solution. At the beginning of the new forward pass, the b-th bee remains loyal to its formerly created solution with the probability

$$p_b^{u+1} = e^{-\frac{1-O_b}{u}}, \qquad b = 1, 2, \ldots, B, \tag{1.4}$$

where O_b represents the normalized value for the current cost of solution formed by the b-th bee, while u keeps information on the number of forward passes ($u = 1, 2, \ldots, NC$). The value of O_b is calculated as follows:

$$O_b = \begin{cases} \frac{C_{max}-C_b}{C_{max}-C_{min}}, & \text{if } C_{max} \neq C_{min}, \\ 1, & \text{if } C_{max} = C_{min}, \end{cases} \qquad b = 1, 2, \ldots, B, \tag{1.5}$$

where C_{min} and C_{max} denote minimum and maximum values over the costs of all solutions currently held by the bees.

Within the RECRUITINGPROCESS procedure, using the roulette wheel method, each follower decides on a new vessel arrangement by choosing one arrangement from the set of the solutions generated by the recruiters. Recruiters with better solutions (smaller current costs) have greater probabilities to be selected by a follower. The probability that the solution of recruiter b would become the solution for any of the uncommitted bees is calculated as:

$$p_b = \frac{O_b}{\sum_{k=1}^{R} O_k}, \qquad b = 1, 2, \ldots, R, \tag{1.6}$$

where O_k denotes the normalized value for the current cost of the k-th recruiter and R represents the number of bees that recruit. Roulette wheel method is applied to assign each uncommitted follower to one recruiter by using (1.6) and a random number generator. Each bee that abandoned its own solution will take the solution of its recruiter bee. More precisely, the recruiter's solution is copied and placed to be the solution of the follower. In addition, the follower copies the Ξ list from the recruiter as well. In case that the solution is not complete, meaning that there are vessels which have not been allocated on the berth-time plane, the algorithm will proceed with a new forward pass.

Forward and backward passes are performed NC times each, such that a complete solution is constructed for each bee. After NC steps, some of B bees managed to create complete (i.e., feasible) solutions, whereas others fail to do that. This failure is due to the fact that both the vessel ordering and their positions are not adequate, as there might be vessels blocking the remaining positions for the other available vessels. Among all feasible solutions, the one with the best total cost is determined. The selected solution is used to update the new global best, meaning that an iteration of BCO is completed. The BCO for MCHBAP is enhanced by the improvement techniques that are described in the remaining part of the section. The BCO iterations and improvement steps are executed until the stopping condition (maximum CPU time) is met. Finally, the best formed solution (i.e., the current global best) is returned as the final BCO solution.

1.4.2 Improvement Techniques of BCO

To enhance the constructive BCO approach, three improvement techniques are proposed in [22] that are performed on the complete solutions.

The first improvement technique is used at the end of each iteration and it is applied to all complete solutions held by the bees. Each bee which has a complete solution tries to accomplish some improvements of its own solution by using the final ξ list. Each bee moves trough its own list of vessels looking for the vessel with the highest penalty cost that has a nonempty cheaper ξ element. Each vessel $v_k, k \in \{1, \ldots, l\}$ is moved to the first (cheapest) available position according to the existing ξ list, if such a position exists. After this step, the Ξ list is updated for all unused vessels indicating the new forbidden positions. Due to the complexity, the clearing of the newly available positions is not performed at this phase.

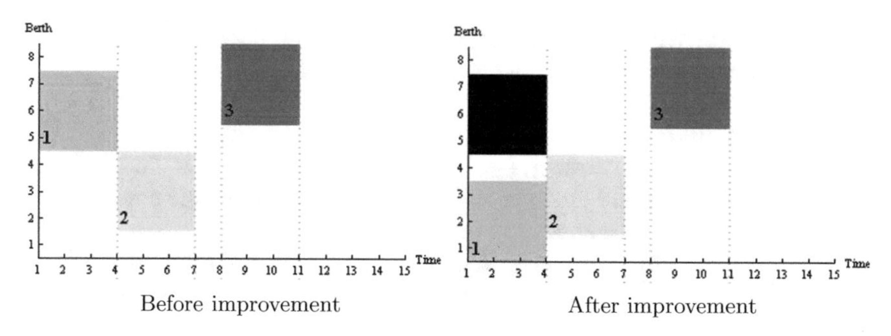

Fig. 1.7 Illustration of the first improvement procedure

The first improvement procedure is performed for each vessel in the the list, so that either a better solution is formed for the bee, or it remains unchanged if all lists for all vessels do not contain cheaper positions. Finally, each solution is being evaluated and the one with the lowest penalty cost is set as the global best solution.

Figure 1.7 shows an example of the vessels' allocations before and after the first improvement of the complete solution. Positions that have not been released are colored in black in Fig. 1.7. After reallocation of the vessel 1, its previously occupied positions remain marked, even they are free for new allocations. However, as Ξ list is not updated, these positions are not considered by other vessels. For example, suppose that position on berth 6 and time unit 3 is cheaper for vessel 3 compared to its currently allocated position. Before allowing the algorithm to move vessel 3 on this cheaper position, the Ξ list must be updated.

Algorithm 6 The first improvement procedure

```
procedure FIRSTIMPROVEMENT(BeeΞ, BeeSol)
   SORT(vessels)
   for i ∈ vessels do
      if EXISTSSMALLERCOST(BeeSol(i)) then
         BeeSol ← UPDATESOL(BeeSol(i), BeeΞ(i, 1))
      end if
   end for
   BESTSOLUPDATE()
end procedure
```

The above described ideas are implemented in the first improvement procedure with pseudo-code presented in Algorithm 6. The procedure FIRSTIMPROVEMENT is called for all bees with complete solutions. The vessels are ordered non-increasingly in terms of the penalty costs by applying the procedure SORT. Procedure EXISTSSMALLERCOST returns *True* if the *Bee*ξ list is containing an element with lower penalty cost than the penalty cost on position p_i for vessel i. If algorithm determines a better position for the examined vessel, solution has to be updated in appropriate way by procedure UPDATESOL. At the end of first improvement, the obtained *Solution* is used to update *GlobalBest* value by procedure BESTSOLUPDATE.

Note that within FIRSTIMPROVEMENT procedure, only the cheapest (first) position from existing ξ list for each vessel v_k, $k = 1, \ldots, l$ has to be examined for possible conflict with some other vessel v_i, $i = 1, \ldots, l$, $i \neq k$. It means that there exist two loops counting vessels and executing $l(l-1)$ times. Therefore, time complexity of FIRSTIMPROVEMENT procedure is $O(l^2)$.

Let us assume that the vessels v_k, $k \in \{k_1, \ldots, k_i\}$ with $1 \leq i \leq l$ (initially allocated to the positions p_k) are now moved to the positions n_k. As all p_k positions are now free, it may happen that there are vessels which can be moved to the positions with lower cost than the current one. Generally, Ξ list consists of many elements and therefore, its frequent update can be time consuming. This is the reason why Ξ list cannot be updated properly within the first improvement. For this reason, we introduce the second improvement procedure, which is applied only to the *Solution* having *GlobalBest* cost.

In order to make positions p_k in *Solution* available for other vessels, the Ξ list needs to be updated in the appropriate way. As this operation would be too expensive, a completely new allocation is created as follows. Vessels v_k, $k \in \{k_1, \ldots, k_i\}$ are allocated to the corresponding position n_k. All other vessels v_k, $k \in \{1, \ldots, l\} \setminus \{k_1, \ldots, k_i\}$ are fixed on previously considered positions p_k. The allocation starts from the initial Ξ list. Vessels are taken one by one, and after the vessel v_k, $k \in \{1, \ldots, l\}$ is allocated to the position n_k or p_k, the ξ lists for all other vessels v_n, $n \neq k$ are cleared. For each vessel v_n, all positions in the corresponding ξ list are checked for possible conflict with the fixed position of vessel v_k and the conflicting positions are deleted. The process is iterated for all vessels and their fixed positions in global best solution. When the process is finished, all vessels on their fixed positions are saved in *Solution* and the Ξ list is properly updated. Hopefully, the new Ξ list will contain the positions with smaller cost for some of the vessels. In this way, an initial solution for new, second improvement procedure of the complete solution, is obtained.

In order to reduce the CPU time, the second improvement is applied only after each *NC* iterations. Once BCO algorithm performs *NC* iterations and *Solution* and *GlobalBest* are updated, new improvement is performed in the following manner. At the beginning, all vessels are arranged in non-increasing order according to the penalty costs in *Solution*. The first vessel (the one with the highest penalty cost) is considered for moving to the position with the smallest possible cost. If the reallocation is performed (at least one position with smaller cost than the current one exists), the complete update of Ξ list is performed, as well as the new sorting of vessels. In the case when the position with smaller cost than the current one does not exist, the next vessel from the sorted list is considered. The algorithm repeats this procedure as long as possible, i.e., as long as it is possible to reallocate at least one vessel.

The pseudo-code for the second improvement is given in Algorithm 7. At the beginning of procedure SECONDIMPROVEMENT, Ξ gets value of the *start*ξ list. Procedure POSSIBLETOMOVE returns a single element set with the index of the leftmost vessel with existing cheaper positions. The task of the procedure FIXANDCLEAN is to move identified vessel to the cheapest available position and completely update Ξ list as well as the corresponding *Solution*.

Algorithm 7 The second improvement procedure

```
procedure SECONDIMPROVEMENT( )
    repeat
        Ξ ← startΞ
        SORT(vessels)
        V_m ← POSSIBLETOMOVE(vessels)
        if V_m ≠ {} then
            FIXANDCLEAN(Ξ)
        end if
    until V_m = {}
    BESTSOLUPDATE()
end procedure
```

As it can be seen from Algorithm 7, procedure SECONDIMPROVEMENT has additional steps for updating ξ lists, which can be done by $l(l-1)$ operations. As in the case of FIRSTIMPROVEMENT procedure, for each vessel first position from ξ list is checked for possible conflict with all other vessels, demanding additional $l(l-1)$ operations. Each vessel is allocated to some current position with some current cost in feasible solution. Subset of cheaper positions from ξ list has to be selected and verified against conflict. In the worst case, the number of cheaper positions is equal to $mT-1$. Consequently, the time complexity of SECONDIMPROVEMENT procedure is $O((l(l-1)+l(l-1)) \cdot (mT-1)) = O(l^2mT)$.

An example of the vessels' allocations before and after second improvement is presented in Fig. 1.8. After reallocation of vessel 1 by first improvement procedure, its previous positions are not available (marked with black color). The second improvement procedure performs update of Ξ list and cheaper position for the vessel 3 is released on berth 6 at time unit 3. Vessel 3 can be now moved on that position, and Ξ list remains properly updated after all performed reallocations.

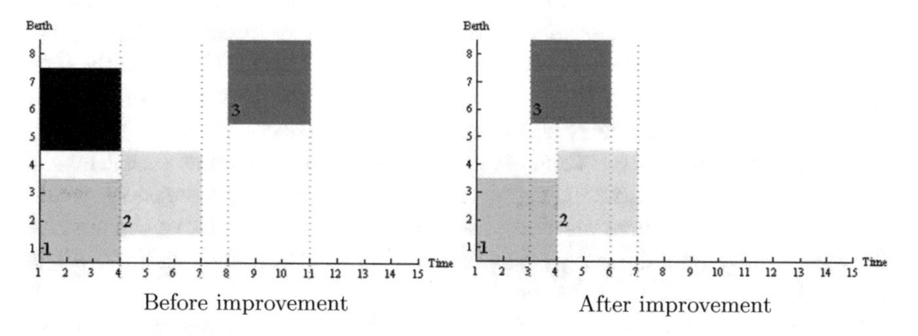

Fig. 1.8 Illustration of the second improvement procedure

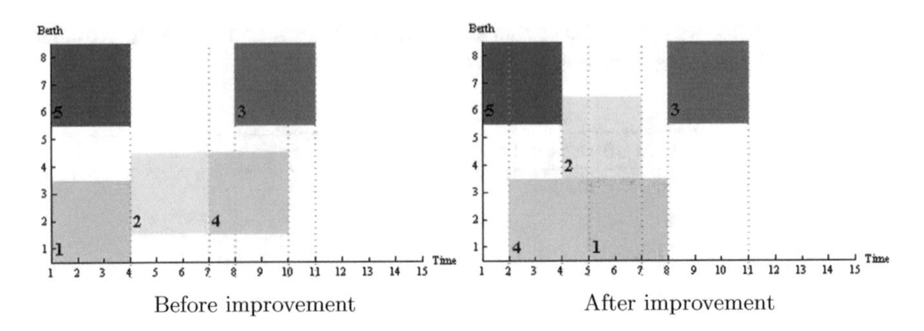

Fig. 1.9 Illustration of the third improvement procedure

At each reduction of the *GlobalBest* value, the best found solution undergoes the third, final improvement. For each vessel, the third improvement considers all unfeasible positions with smaller penalty costs than the current one. The infeasibility means that those positions are occupied by other vessels. We would refer to those vessels as conflicting with the considered vessel. The third improvement tries to resolve the conflict in such a way that the reallocation of conflicting vessels yields the reduction of total cost.

An illustration of the third improvement technique is presented in Fig. 1.9. For example, let us assume that for vessel 4, the position at berth 1 and time unit 2 is cheaper than its current one. This position is already occupied by vessel 1. If this position is taken into account as possible allocation for vessel 4, it would be in conflict with vessels 1 and 2. Thus, the third improvement procedure tries to reorganize vessels 4, 1, and 2, such that the overall cumulative cost is minimized. Small shift of vessels 1 and 2 allows that vessel 4 can be allocated to that cheaper position.

The structure of the third improvement technique is presented in Algorithm 8. As in previous two improvement techniques, vessels are arranged in the decreasing order in respect to the penalty costs. Some vessels were obviously not allocated to cheaper positions due to the conflicting reasons. The idea of the third improvement is to enable considering these positions and resolving conflicts by repositioning some other vessels. For a given vessel i and all its possible cheaper positions j, procedure CONFLICTVESSELS(i) creates the subsets of conflicting vessels V_j. Generally, there are few vessels conflicted with the vessel allocated at the observed position, and therefore, this search does not lead to large CPU time consumption.

For each position j, vessel i and set V_j are considered at the same time. The idea is to slightly shift the conflicting set of vessels V_j, such that the increase of penalty costs of vessels V_j produces the significant decrease in the cumulative penalty cost. This means that the total penalty cost of the vessel i and the set of vessels V_j will be decreased with respect to the *Solution*.

The role of the function SOLVE(s, Ξ) is to find the new positions for vessels from the given list s by examining all available positions in Ξ list. If some vessel is moved from its original position, the obtained new arrangement of vessels is saved

Algorithm 8 The third improvement procedure

```
procedure ThirdImprovement( )
    repeat
        Sort(vessels)
        temp ← GlobalBest
        for i ∈ vessels do
            V ← ConflictVessels(i)
            for all V_j do
                movingVessels ← V_j ⋃{i}
                ξ(V_j) ← startΞ(V_j)
                tempAllocation(movingVessels) ← Solve(movingVessels, Ξ)
            end for
            MaxSavings(tempAllocation)
            if Solution ≠ {} then
                BestSolUpdate()
                Break()
            end if
        end for
    until temp = GlobalBest
end procedure
```

in *tempAllocation* variable. Otherwise, *tempAllocation* contains empty set. When all possible cheaper positions are considered, the best of them (yielding the maximal reducing of total cost) is identified by a function MaxSavings. This function also performs the rearrangements of vessels that yields to the improvement, if it exists, or returns empty *Solution*. The variable *Solution* is used to update *GlobalBest*. Third improvement is repeated until there is no change on *GlobalBest*.

In the third improvement procedure, for each vessel v_k, $k = 1, \ldots, l$, the sublist of positions cheaper than the current position of v_k has to be determined. For all cheaper positions, the set of conflicting vessels with v_k has to be formed, which can be done by $l(l-1)(mT-1)$ operations. In the worst case, the number of cheaper positions is $mT-1$ and each vessel can be in conflict with all remaining $l-1$ vessels. Consequently, the algorithm has to choose l positions out of $mT-1$ possible positions for vessel allocation. There are $l!\binom{mT-1}{l}$ different allocations. Bringing together all requests yields the time complexity of $O(l(l-1)(mT-1) + l!\binom{mT-1}{l}) = O(l^2 mT + l!\binom{mT-1}{l})$ for a single pass of the third improvement procedure.

The pseudo-code of the BCO algorithm is shown in Algorithm 9. The variable *run* is used to count iterations of BCO algorithm. After constructive BCO iteration is completed, by calling procedure Iteration, the first improvement is applied to all complete solutions generated by bees. The second improvement is applied only to *GlobalBest* solution after every *NC* iterations. The *GlobalBest* is subject to the third improvement each time it is changed (the total cost is reduced). This means that the third improvement may be applied immediately after the first one. The procedure BCOforMCHBAP runs until stopping criterion defined by maximum run time is met.

Algorithm 9 Bee Colony Optimization algorithm

```
procedure BCOFORMCHBAP(B, NC, RunTime)
    run ← 1
    GlobalBest ← ∞
    while SESSIONTIME() ≤ RunTime do
        ITERATION(B, NC)
        for b ← 1, B do
            FIRSTIMPROVEMENT(BeeΞ(b), BeeSol(b))
        end for
        if MOD(run, NC) = 0 then
            SECONDIMPROVEMENT()
        end if
        if CHANGED(GlobalBest) then
            THIRDIMPROVEMENT()
        end if
        run ← run + 1
    end while
end procedure
```

1.5 Variable Neighborhood Search Based Metaheuristic for MCHBAP

In this section, we first provide description of the Variable Neighborhood Descent method (VND) introduced in [7] for MCHBAP. We further propose General Variable Neighborhood (GVNS) as a new metaheuristic method for MCHBAP. The proposed GVNS involves shaking phase and uses VND over the six neighborhood structures as a local search part.

1.5.1 Variable Neighborhood Descent for MCHBAP

The first implementation of the Variable Neighborhood Descent (VND) method for MCHBAP is proposed in [7]. The MCHBAP solutions in VND are represented by *sequence pair* introduced in [30]. A sequence pair consists of two permutations, H and V, which are used to describe the positions of vessels in the planning horizon. The rules for defining permutations H and V are as follows:

(a) if vessel j precedes vessel i in the permutation H, then vessel j "cannot see" vessel i on "left-up" view,
(b) if vessel j precedes vessel i in the permutation V, then j "cannot see" i on "left-down" view.

The definition of H and V permutations is illustrated in Fig. 1.10. The gray region represents left-up (left-down) view of vessel j, respectively. Therefore, vessel j cannot see vessel i in both cases, and consequently, j is before i in both permutations H and V. Note that the illustrations in Fig. 1.10 do not correspond to a single (H, V) pair.

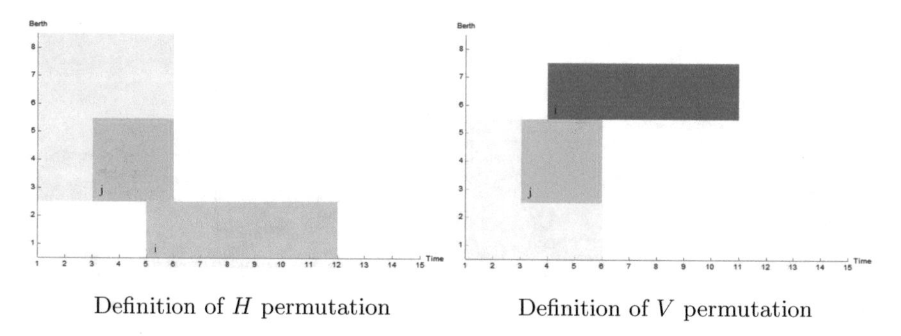

Fig. 1.10 An illustration of sequence pair solution representation based on (H, V) permutations

Actually, according to the left part of Fig. 1.10, j precedes i in permutation H, but it is after i in permutation V. Similarly, j is after i in permutation H and precedes it in V, according to the right part of Fig. 1.10.

Transformation from an allocation to (H, V) pair is unique. On the other hand, a pair of permutations (H, V) represents a whole class of allocations. Therefore, a special procedure is designed to determine an allocation with the minimum total cost. In [30], pair (H, V) represents a solution of VLSI layout design problem with the goal to minimize the total space used. This property is explored to break the ties when decoding (H, V) in [30].

To ensure that permutation pair (H, V) will result in a feasible solution to BAP, the study [7] uses the concept of the *longest common subsequence* (LCS) introduced in [38]. Sequence pair is in connection with weighted directed acyclic graphs that can be assigned to the observed packing problem. These graphs represent horizontal and vertical relationship among elements to be packed. Horizontal path in the packing is a common subsequence of a sequence pair, and vice versa. Vertical path in the packing is defined similarly. The only difference is that the first element of the sequence pair is reversed. The longest (weighted) common subsequences in horizontal and vertical relationship graphs correspond to the longest weighted paths in those graphs [38]. Intuitively, LCSs may be considered as the maximal dimensions required for packing the given elements.

The LCS concept applied to MCHBAP observes permutations H and V as weighted sequences and finds the longest common subsequence in the weighted permutation pair. In [7], LCS is used in combination with vessel size to check the feasibility of the H and V permutations. To check the feasibility with respect to the total number of available berths, the number of used berths is considered as the length of a vessel. In the same way, vessel's width (defined as the required processing time) is used to check the feasibility of the pair (H, V) with respect to the planning horizon. In other words, pair (H, V) leads to a feasible allocation in the two dimensional berth-time plane, if two conditions are satisfied:

$$\sum_{v_j \in LCS(H,V)} \left\lceil \frac{a_j}{b_j} \right\rceil \leq T, \tag{1.7}$$

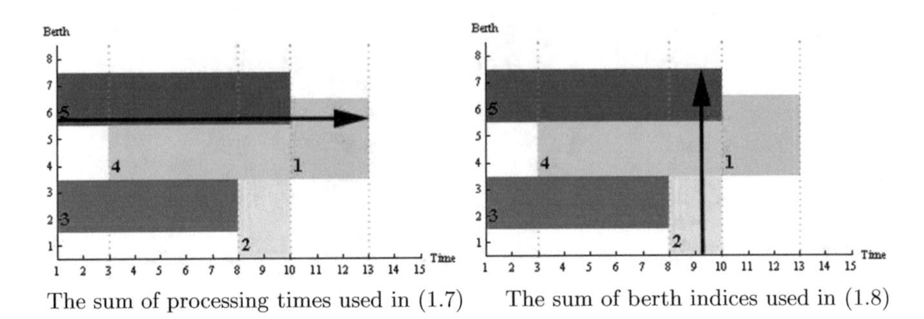

The sum of processing times used in (1.7) The sum of berth indices used in (1.8)

Fig. 1.11 An illustration of LCS concept

$$\sum_{v_j \in LCS(H^R, V)} b_j \leq m, \tag{1.8}$$

where H^R denotes the reverse of permutation H. Pair (H,V) satisfying conditions (1.7) and (1.8) is named *feasible pair* (H, V).

An illustration of feasible pair $(H, V) = (\{5, 4, 3, 2, 1\}, \{3, 2, 4, 5, 1\})$ and the corresponding longest common subsequences with respect to the time and berth axes is given in Fig. 1.11. LCS with respect to time axis is $\{5, 1\}$, while LCS with respect to berth axis is $\{2, 4, 5\}$, marked by the horizontal and vertical black arrows, respectively.

Having in mind that the goal of MCHBAP is to minimize the total cost of allocating berths to vessels, the decoding procedure has to differ from the one described in [30]. Such a procedure, which finds the best reference point corresponding to the given permutations H and V, was introduced in [7].

For two vessels v_i and v_j, binary relations $\lhd$ on the permutation H, as well as on the permutation V, are denoted as $\lhd_H$ and $\lhd_V$, respectively, and they are defined as follows:

$$v_i \lhd_H v_j \Leftrightarrow v_i \text{ precedes vessel } v_j \text{ in permutation } H,$$
$$v_i \lhd_V v_j \Leftrightarrow v_i \text{ precedes vessel } v_j \text{ in permutation } V.$$

For the given set of vessels $\upsilon = \{v_1, v_2, \ldots, v_l\}$ and vessel $v_j \in \upsilon$, disjoint subsets L, R, A and B of υ, based on pair (H,V) are defined as:

$$L(v_j) = \{v_i \mid v_i \lhd_H v_j \wedge v_i \lhd_V v_j \wedge v_i \in \upsilon\}, \tag{1.9}$$

$$R(v_j) = \{v_i \mid v_j \lhd_H v_i \wedge v_j \lhd_V v_i \wedge v_i \in \upsilon\}, \tag{1.10}$$

$$A(v_j) = \{v_i \mid v_i \lhd_H v_j \wedge v_j \lhd_V v_i \wedge v_i \in \upsilon\}, \tag{1.11}$$

$$B(v_j) = \{v_i \mid v_j \lhd_H v_i \wedge v_i \lhd_V v_j \wedge v_i \in \upsilon\}. \tag{1.12}$$

Set L consists of all vessels that have to be allocated to the left from the given vessel v_j. Vessels from the set R have to be placed to the right from the vessel v_j. Sets A and B represents the sets of vessels whose allocation point will be above or below from the allocation point of the given vessel v_j.

Subsets R and A are used to decode feasible pair (H, V) in such a way that total allocation cost is minimized. Namely, algorithm starts from the vessel v_k with empty R and A subsets, and finds its cheapest available position. Such unique vessel exists, it is the rightmost and highest allocated vessel in the port. The cheapest position for vessel v_k is the first element of its ξ list consistent with (H, V) pair. After v_k is allocated, proper update of the ξ lists for all other vessels is performed and vessel v_k is removed from the subsets R and A, ensuring that there will be no overlap for vessels that are still unallocated. Next vessel with empty sets R and A is chosen. For the selected vessel v_j, the algorithm first determines all positions from its ξ list that are not violating order defined by the pair (H, V), and among those positions takes the cheapest one. That position defines the reference point of vessel v_j. One by one, all unallocated vessels are placed in the port. Note that each vessel has at least one feasible allocation point in the port, as the pair (H, V) is also feasible. Procedure of decoding pair (H, V) is described in Algorithm 10.

Algorithm 10 Decoding of pair (H, V)

```
procedure DECODE(H, V, vessels)
  if FEASIBLE(LCS(H,V)) then
    for v_i ∈ vessels do
      {R(v_i), A(v_i)} ← EXAMINE(H, V)
    end for
    notAllocated ← vessels
    Ξ ← startΞ
    while notAllocated ≠ {} do
      v_j ← FINDEMPTY(R, A)
      positions ← CLEAR(ξ(v_j))
      Solution(v_j) ← FINDCHEAPEST(positions)
      notAllocated ← notAllocated\{v_j}
      Ξ ← UPDATELIST(Ξ)
    end while
  else
    Solution ← {}
  end if
end procedure
```

Procedure EXAMINE forms the subsets of vessels R and A for each element from the set of observed vessels, while function FINDEMPTY returns the index of vessel with empty subsets R and A. To ensure that only feasible positions that are in accordance with pair (H, V) are used for vessel allocation, the algorithm calls procedure CLEAR. Among positions returned by CLEAR procedure, function FINDCHEAPEST selects the best one, i.e., position with the smallest penalty cost. After allocating a vessel in the port, procedure UPDATELIST is called to update all ξ lists.

The VND proposed in [7] starts with the procedure INITIALSOLUTION that constructs an initial solution. At the beginning, procedure CLUSTERPREFERREDLOCATIONS is applied to generate groups of vessels that are in conflict regarding their preferred berthing locations. This step is followed by procedure SORT, which is sorting the obtained groups in non-increasing order of their cardinality. Within a group, vessels are sorted in non-decreasing order based on the values of parameter ETA. One by one, groups of vessels are allocated in the port by procedure ALLOCATE, in such a way that the total cost of a group is minimized. Having in mind that the groups of conflicting vessels are usually small, all possible ordering of vessels are examined for scheduling within the available positions in their ξ lists, and the one resulting in minimum total cost is selected for allocation. If a vessel is not in a conflict with any other vessel, it is allocated on the very first (the cheapest) position of its ξ list. This step is denoted in pseudo-code by $Solution(groups(i)) \longleftarrow \xi(groups(i), 1)$. After a proper position for a vessel is found, the complete update of Ξ list elements for all remaining vessels is performed by procedure UPDATE. When all vessels are allocated (i.e., all groups are processed) an initial solution is formed. The pair of permutations (H, V) corresponding to the initial solution is formed. The pseudo-code of the described procedure is shown in Algorithm 11.

Algorithm 11 Construction of the initial solution for VND

```
procedure INITIALSOLUTION(vessels)
   groups ← CLUSTERPREFERREDLOCATIONS(vessels)
   groups ← SORT(groups)
   i ← 1
   Ξ ← startΞ
   while i ≠ LENGTH(groups) do
      if LENGTH(groups(i)) ≠ 1 then
         Solution(groups(i)) ← ALLOCATE(groups(i), Ξ)
      else
         Solution(groups(i)) ← Ξ(groups(i), 1)
      end if
      Ξ ← UPDATE(Ξ)
      i ← i + 1
   end while
   RETURN (groups)
end procedure
```

The obtained initial feasible solution is then represented by the corresponding initial permutations H and V. In addition, the vessels that are not allocated to their most preferred positions are identified and grouped in a set ωS. Within ωS, the vessels are sorted in non-increasing order based on the costs corresponding to the current best solution. Each time ωS is modified, the sorting of its elements is performed.

By comparing H and V permutations corresponding to the initial and the best known solutions of several test instances, it was noticed that the vessels missing their most preferred positions (i.e., the vessels from the group ωS) should be moved

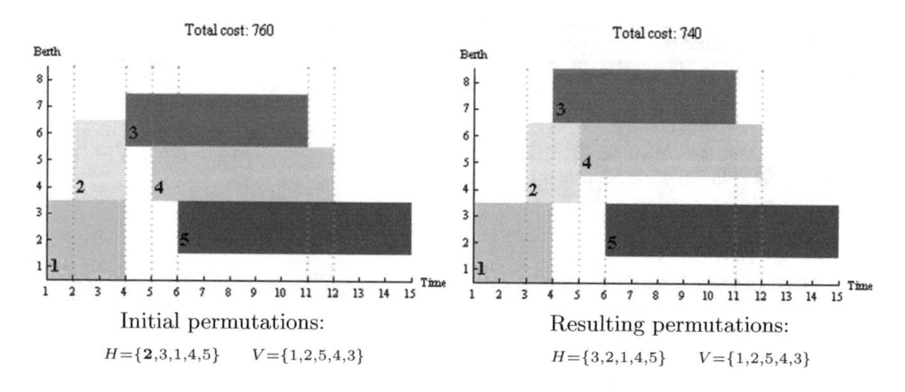

Fig. 1.12 The result of applying *ChangePositionH* to vessel 2 that is moved $k = 1$ positions to the right in respect to permutaion H

only few positions to the left or right in both initial permutations H and V. This was the main idea for defining neighborhood structures in the VND algorithm from [7].

The VND proposed in [7] uses three types of neighborhoods that are applied only to the vessels from ωS, respecting the order of vessels. For a given size k, $k = 1, 2, 3, \ldots, k_{max}$, the three types of neighborhoods are applied as follows:

(i) *ChangePositionH* is applied first. The selected vessel is moved k positions to the left in permutation H. In the case when improvement occurs, changed permutation H becomes a starting point of a new local search and k is set to 1. If no improvement is obtained, the same vessel is moved k positions to the right in H, while permutation V remains unchanged;

(ii) *ChangePositionV* is used next. The selected vessel is first moved k positions to the left in permutation V. The idea used is the same as the technique applied in the first neighborhood, if an improvement is achieved. If no improvement occurs, the same vessel is moved k positions to the right in permutation V;

(iii) *ChangePositionHV* is the last to be applied. This neighborhood type is a combination of *ChangePositionH* and *ChangePositionV*. In the neighborhood *ChangePositionHV*, all possible changes of H and V are examined.

Examples of *ChangePositionH*, *ChangePositionV*, and *ChangePosition HV* for $k = 1$ are illustrated in Figs. 1.12, 1.13 and 1.14, respectively. It is obvious that parameter k_{max} may take values between 1 and $l-1$, however, its actual range depends on the current position of the considered vessel. Pseudo-code of the VND from [7] is presented in Algorithm 12.

Algorithm 12 Variable Neighborhood Descent algorithm

```
procedure VND(vessels, k_max, RunTime)
    INITIALSOLUTION(vessels)
    {H, V} ← PERMUTATIONS()
    ωS ← NOTPREFERREDPOSITION()
    k ← 1
    while k ≤ k_max do
        i ← 1
        noImprovement ← True
        while i ≤ LENGTH(ωS) ∧ noImprovement do
            H ← CHANGEPOSITIONH(ωS(i), k)
            temp ← DECODE(H, V, vessels)
            if CALCULATECOST(temp) < GlobalBest then
                ωS ← NOTPREFERREDPOSITION()
                k ← 1
                noImprovement ← False
                BESTSOLUPDATE()
            else
                i ← i + 1
            end if
        end while
        if noImprovement then
            i ← 1
        end if
        while i ≤ LENGTH(ωS) ∧ noImprovement do
            V ← CHANGEPOSITIONV(ωS(i), k)
            temp ← DECODE(H, V, vessels)
            if CALCULATECOST(temp) < GlobalBest then
                ωS ← NOTPREFERREDPOSITION()
                k ← 1
                noImprovement ← False
                BESTSOLUPDATE()
            else
                i ← i + 1
            end if
        end while
        if noImprovement then
            i ← 1
        end if
        while i ≤ LENGTH(ωS) ∧ noImprovement do
            {H, V} ← CHANGEPOSITIONHV(ωS(i), k)
            temp ← DECODE(H, V, vessels)
            if CALCULATECOST(temp) < GlobalBest then
                ωS ← NOTPREFERREDPOSITION()
                k ← 1
                noImprovement ← False
                BESTSOLUPDATE()
            else
                i ← i + 1
            end if
        end while
        if noImprovement then
            k ← k + 1
        end if
    end while
end procedure
```

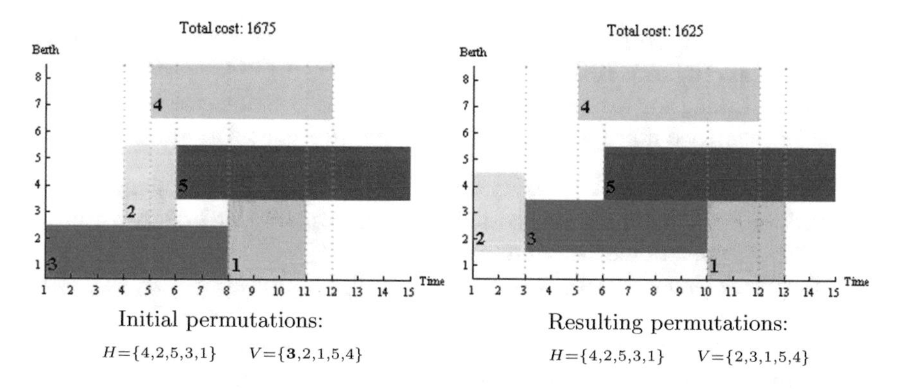

Fig. 1.13 The result of applying *ChangePositionV* to vessel 3 that is moved $k = 1$ positions to the right in respect to permutation V

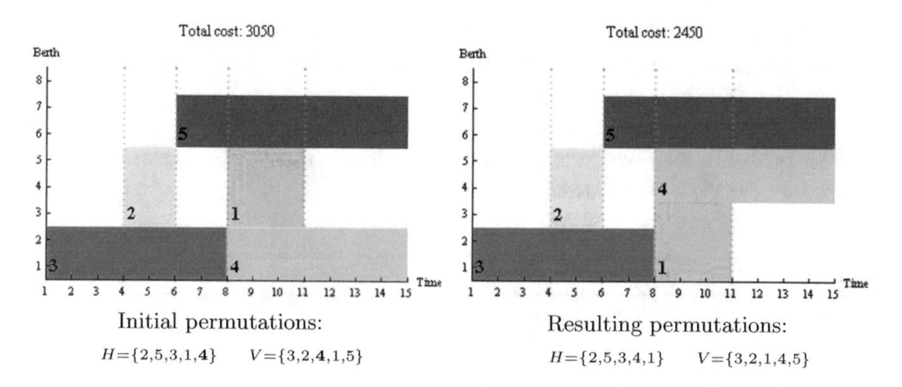

Fig. 1.14 The result of applying *ChangePositionHV* to vessel 4 that is moved $k = 1$ positions to the left with respect to permutation H and $k = 1$ positions to the right in respect to permutation V

1.5.2 General Variable Neighborhood Search for MCHBAP

To improve the performance of the VND from [7] and to allow better diversification of results, a General Variable Neighborhood Search (GVNS) for MCHBAP is developed. The proposed GVNS involves shaking phase and uses six neighborhoods in VND. Therefore, this VND is different from the one proposed in [7]. Shaking procedure involves stochastic transformation of current best solution, and it is used in order to help the algorithm to escape from the local minimum and to efficiently explore the search space. In the local search phase, instead of simple local search, the proposed GVNS uses VND to find improvements by systematic exploration of six neighborhoods of the current solution.

The proposed GVNS starts with the procedure INITIALSOLUTION described in Algorithm 11. The initial solution is generated in the same way as in VND. The further execution of GVNS is based on the groups of conflicted vessels, which is not the case in VND. After the initial solution is generated, GVNS performs the two main steps (shaking and local search) until the stopping criterion is satisfied. Within the shaking step, two transformations are performed to obtain a new solution:

(i) First, k random groups of vessels are selected, based on the calculated priority that is proportional to the total cost of the group. The selected groups are moved at the beginning of the list of groups;
(ii) After that, k random pairs of vessels are selected, based on the calculated priority that is proportional to the vessel's cost in the current best solution. The positions of the selected pairs of vessels are exchanged. For this transformation, it is not important whether vessels belong to the same group or not.

The structure of shaking phase is represented in Algorithm 13. The role of CALCULATECOST procedure is to evaluate the total cost for group of vessels or for a vessel, based on the vessel's cost in the current best solution. Procedure SELECTGROUP selects one group of vessels from the list of groups, based on the calculated priority. The selected group is moved as the first element in the list of groups by procedure PUTASFIRST. Similarly, procedure SELECT2VESSELS pics two vessels based on their probabilities. Two selected vessels exchange their positions in the list of groups by procedure EXCHANGEPOSITIONS.

Algorithm 13 Shaking phase in GVNS

```
procedure SHAKE(vessels, groups, k)
    done ← False
    while ¬done do
        p ← CALCULATECOST(groups)
        for i ← 1, k do
            g ← SELECTGROUP(groups, p)
            groups ← PUTASFIRST(groups, g)
        end for
        p ← CALCULATECOST(vessels)
        for i ← 1, k do
            {v1, v2} ← SELECT2VESSELS(vessels, p)
            groups ← EXCHANGEPOSITIONS(groups, v1, v2)
        end for
        ShakeSol ← ALLOCATE(groups)
        done ← FEASIBLE(ShakeSol) ∧ NEW(ShakeSol)
    end while
    {H, V} ← SHAKEPERM(ShakeSol)
    RETURN (H, V, groups)
end procedure
```

The newly formed groups are allocated following the same rules as in the procedure for constructing the initial solution. New solution is formed by procedure ALLOCATE that uses the list of newly formed groups as input. If algorithm determines that the formed vessels clustering leads to an unfeasible solution, shaking procedure of the current best solution is repeated until feasible solution is constructed. If the newly created solution is already examined during algorithm's run, shaking procedure of the current best solution is performed again. Procedure FEASIBLE investigates whether the newly formed solution is feasible or not, while the procedure NEW checks if the newly formed solution has already been explored during the algorithm's run. Once the shaking procedure creates a new feasible solution, the corresponding permutation pair (H, V) is constructed by SHAKEPERM procedure. After that, the GVNS algorithm proceeds to the local search phase that is performed on the new feasible solution generated by the procedure SHAKE.

In the local search phase, the proposed GVNS uses VND instead of simple local search to find improvements of the current solution. Six neighborhoods are used to systematically explore solution space. These neighborhoods are changing H or V permutation, or both permutations simultaneously, in order to form the optimal sequence pair. Sufficiently large neighborhood and appropriate order of permutations transformation are essential in creating high quality solutions. Modifications of standard *Swap* and *Move* are used to reshuffle permutations H and V, and they are applied in the following order:

(i) *SingleSwapH* selects two vessels and exchanges their positions in permutation H, leaving permutation V unmodified;
(ii) *SingleSwapV* selects two vessels and swaps their positions in permutation V, while permutation H is unchanged;
(iii) *SingleMoveH* selects two vessels v_i and v_j and moves vessel v_j behind vessel v_i in permutation H, regardless of mutual order of vessels v_i and v_j;
(iv) *SingleMoveV* performs moving of selected vessel v_j behind vessel v_i in permutation V, regardless of whether the vessel v_i is in front of or behind vessel v_j in the current permutation;
(v) *DoubleSwapHV* consists of one *SingleSwapH* and one *SingleSwapV* that are not necessarily performed on the same pair of vessels;
(vi) *DoubleMoveHV* is performed by one *SingleMoveH* and one *SingleMoveV* that are not necessarily realized on the same pair of vessels.

The described modifications of *Swap* and *Move* procedures are illustrated in Figs. 1.15, 1.16, 1.17, 1.18, 1.19, and 1.20. The pseudo-codes of procedures *Single SwapH*, *SingleMoveV*, and *DoubleMoveHV* are presented in Algorithms 14–16, respectively, while remaining procedures have similar structure.

As it is shown in Algorithm 14, SINGLESWAPH calls a procedure SWAPH that performs a swap modification of the current H permutation *tempH*. This procedure also tracks if there exists a feasible swap transformation not already applied to *tempH*, and returns this information in the variable *ExistsTransformation*. If a new best solution is found, appropriate updates of global variables are performed. At the same time, counter k is set on 1, and exploration of the new best solution continues in the first neighborhood.

In the proposed GVNS, neighborhoods are changing in the order as they are listed. GVNS incorporates the *first improvement* principle. More precisely, when

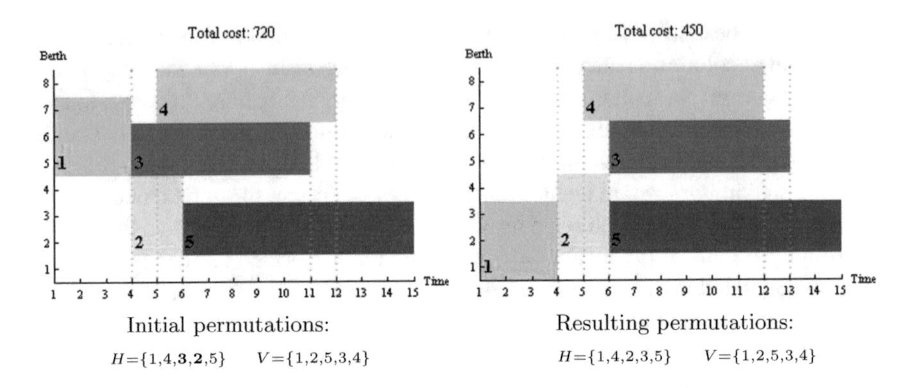

Fig. 1.15 Illustration of *SingleSwapH*

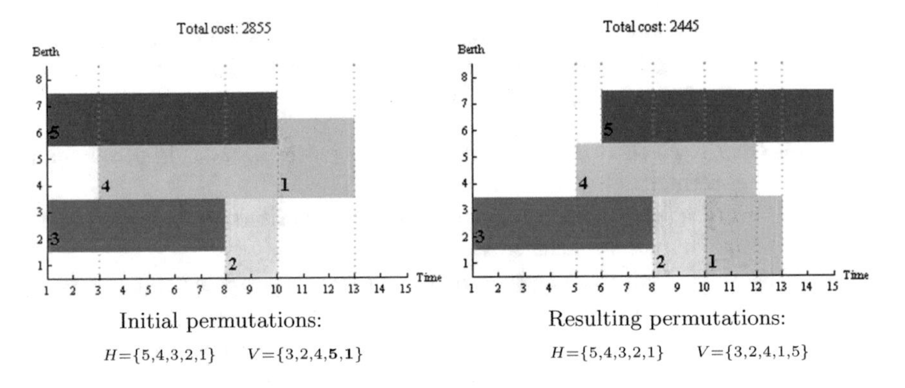

Fig. 1.16 Illustration of *SingleSwapV*

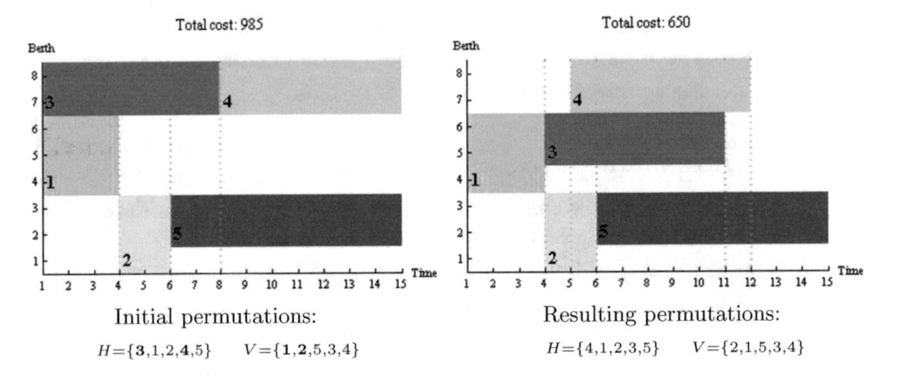

Fig. 1.17 Illustration of *DoubleSwapHV*

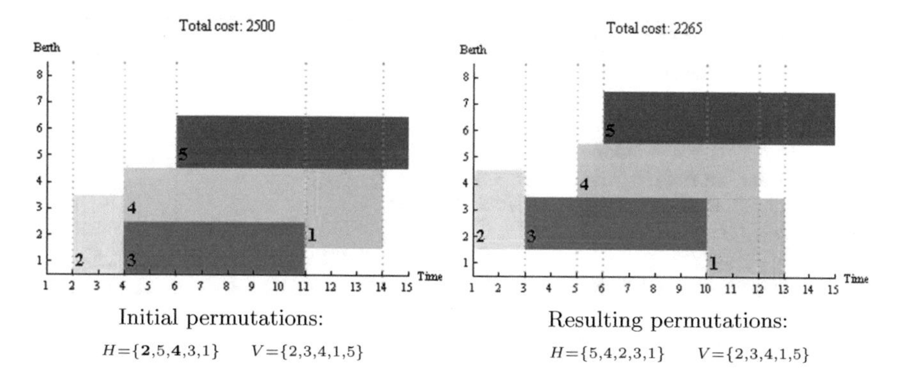

Fig. 1.18 Illustration of *SingleMoveH*

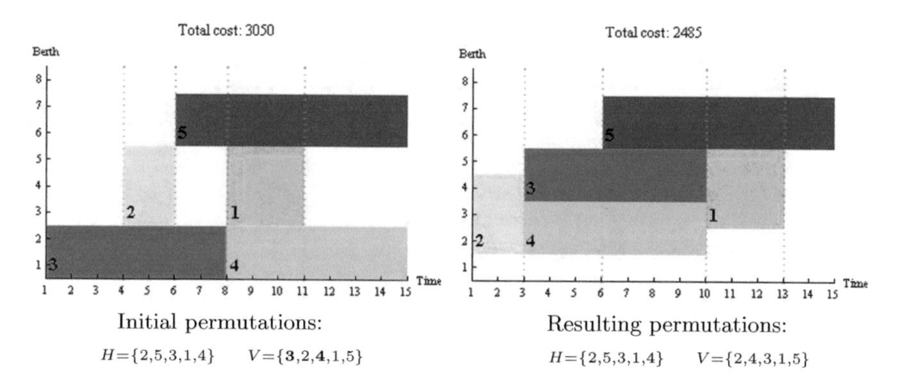

Fig. 1.19 Illustration of *SingleMoveV*

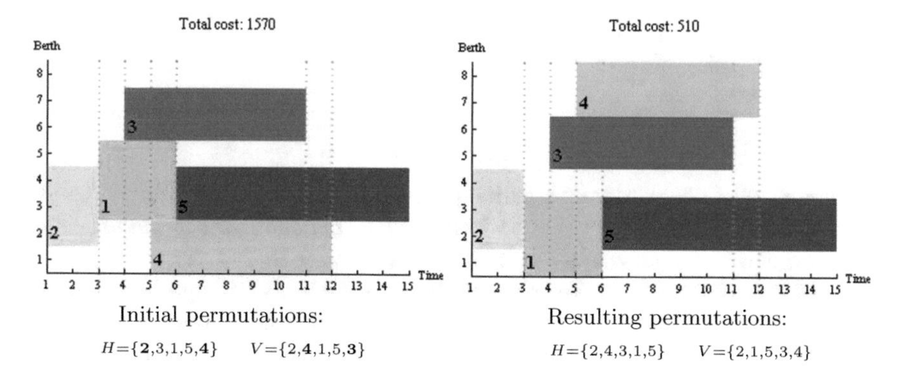

Fig. 1.20 Illustration of *DoubleMoveHV*

Algorithm 14 Procedure *SingleSwapH*

```
procedure SINGLESWAPH(noImprovement)
    ExistsTransformation ← True
    while ExistsTransformation ∧ noImprovement do
        {H_1, ExistsTransformation} ← SWAPH(tempH)
        temp ← DECODE(H_1, tempV, vessels)
        if CALCULATECOST(temp) < GlobalBest then
            H ← H_1
            V ← tempV
            groups ← tempGroups
            k ← 1
            noImprovement ← False
            BESTSOLUPDATE()
        end if
    end while
    RETURN (noImprovement)
end procedure
```

Algorithm 15 Procedure *SingleMoveV*

```
procedure SINGLEMOVEV(noImprovement)
    ExistsTransformation ← True
    while ExistsTransformation ∧ noImprovement do
        {V_1, ExistsTransformation} ← MOVEV(tempV)
        temp ← DECODE(tempH, V_1, vessels)
        if CALCULATECOST(temp) < GlobalBest then
            V ← V_1
            H ← tempH
            groups ← tempGroups
            k ← 1
            noImprovement ← False
            BESTSOLUPDATE()
        end if
    end while
    RETURN (noImprovement)
end procedure
```

an improvement of the current solution is found in one of the six neighborhoods, the neighborhood counter is reset to 1 and the search continues in *SingleSwapH* neighborhood of the new solution. The value of parameter k_{max}, representing the maximal neighborhood size for shaking, is set to $\lceil \frac{l}{2} \rceil$. GVNS algorithm finishes when the stopping criterion is satisfied, i.e., when predefined amount of CPU time is used. Pseudo-code of the described GVNS is represented in Algorithm 17.

Algorithm 16 Procedure *DoubleMoveHV*

```
procedure DOUBLEMOVEHV(noImprovement)
    ExistsTransformation ← True
    while ExistsTransformation ∧ noImprovement do
        {H1, V1, ExistsTransformation} ← MOVEHV(tempH, tempV)
        temp ← DECODE(H1, V1, vessels)
        if CALCULATECOST(temp) < GlobalBest then
            H ← H1
            V ← V1
            groups ← tempGroups
            k ← 1
            noImprovement ← False
            BESTSOLUPDATE()
        end if
    end while
    RETURN (noImprovement)
end procedure
```

Algorithm 17 General Variable Neighborhood Search algorithm

```
procedure GVNS(vessels, k_max, RunTime)
    {groups} ← INITIALSOLUTION(vessels)
    {H, V} ← PERMUTATIONS()
    k ← 1
    while SESSIONTIME() ≤ RunTime do
        {tempH, tempV, tempGroups} ← SHAKE(vessels, groups, k)
        noImprovement ← True
        noImprovement ← SINGLESWAPH(noImprovement)
        noImprovement ← SINGLESWAPV(noImprovement)
        noImprovement ← SINGLEMOVEH(noImprovement)
        noImprovement ← SINGLEMOVEV(noImprovement)
        noImprovement ← DOUBLESWAPHV(noImprovement)
        noImprovement ← DOUBLEMOVEHV(noImprovement)
        if noImprovement then
            k ← k + 1
            if k > k_max then
                k ← 1
            end if
        end if
    end while
end procedure
```

1.6 Experimental Results

Metaheuristics EA [24], BCO [22], VND [7], and the newly introduced GVNS developed for MCHBAP are evaluated and compared against each other and against exact solver from [21]. The computational study is conduced over two sets of problem instances. The first data set contains real-life instances, which are derived from the example used in [4] characterized by $l = 21$ vessels, $m = 12$ berths, and the time

Table 1.2 Vessel specifications for generated test instances

Vessel type	Population (%)	Time range	C1	C2	C3	C4	NBerths
Feeder	60	1–3	2	3	3	9	1
Medium	30	4–5	3	6	6	18	2
Mega	10	6–8	4	9	9	27	3

horizon of $T = 54$ units. For the purpose of our computational experiments, the example from [4] is modified by including new vessels, resulting in 28 vessels in total.

The second data set contains randomly generated instances involving $l = 35$ vessel in the case of $m = 8$ berths, and $T = 112$ time units. These examples proved to be extremely hard to solve, having in mind that exact solver [21] was not able to find their optimal solutions. The set of test instances involves three types of vessels: *feeder*, *medium*, and *mega*. The types of vessels are specified in accordance to the data given in Table 1.2 [27]. For each vessel type, Table 1.2 shows the following data: the percentage of test instances, handling time range, penalty costs (in units of US$ 1000), and the number of berths occupied by specific type of vessel. For the generated instances, the distribution of the favored berthing positions for vessels is homogeneous.

The exact solver from [21] and all four metaheuristic approaches are coded in the *Wolfram Mathematica v8.0*. It is important to note that, unlike classical programming languages, *Mathematica* interprets instructions and therefore, the running times of algorithms may increase. However, our comparison is fair, as all algorithms are executed under the same conditions. All computational experiments with EA, BCO, VND, GVNS, and exact solver from [21] were conducted on the same platform with *Intel Pentium 4* 3.00-GHz CPU and 512 MB of RAM under *Microsoft Windows XP Professional Version 2002 Service Pack 2* OS. Having in mind that metaheuristics are stochastic methods, their stability is examined through repeated runs on each instance. For this reason, BCO, EA, and GVNS methods were run 10 times with time limit of 10 min for all test examples. VND is deterministic in nature, and therefore, it was run only once on each tested instance.

The comparison of results on real-life test examples is presented in Table 1.3. The first column of Table 1.3 contains the number of vessels for each instance from the real-life data set. The objective function value (cost) of the optimal solution obtained by exact solver from [21] is given in the second column named *Opt*, while the corresponding running time (in seconds) is presented in column denoted by T. The next two columns in Table 1.3 contain results obtained by Evolutionary algorithm (EA): the average value over the best found total costs obtained through 10 EA runs ($AvgC$) and the corresponding average minimum CPU time in which EA reaches its best solution for the first time ($AvgT$). The results of BCO and GVNS are presented in the same manner as the results of EA. As VND is deterministic in nature, two columns related to VND contain the best cost obtained in a single VND run (*Best*)

Table 1.3 Computational results—real-life test examples: $m = 12$, $T = 54$

l	Exact solver		EA		BCO		VND		GVNS	
	Opt	*T*	*AvgC*	*AvgT*	*AvgC*	*AvgT*	*Best*	*T*	*AvgC*	*AvgT*
21	4779	4.22	4779	14.86	4779	11.43	4779	0.50	4779	**0.08**
22	4983	6.14	4983	21.10	4983	30.82	4983	0.55	4983	**0.08**
23	5193	8.14	5193	24.83	5193	22.51	5193	0.61	5193	**0.09**
24	5643	8.43	5643	31.25	5643	21.95	5643	0.66	5643	**0.09**
25	5953	15.00	5953	36.47	5953	47.45	5953	0.67	5953	**0.10**
26	6298	64.88	6298	33.72	6298	43.93	6298	0.83	6298	**0.12**
27	6478	66.28	6478	41.12	6478	56.46	6478	0.89	6478	**0.13**
28	6980	326.77	6980	67.06	6980	186.87	6980	0.95	6980	**0.14**
Average	5788.4	62.48	5788.4	33.80	5788.4	52.68	5788.4	0.71	5788.4	**0.10**

and the corresponding VND running time (T). All time related data are given in seconds.

From the results presented in Table 1.3, it can be seen that all four metaheuristic were able to produce optimal solutions provided by exact solver. Moreover, the optimal solution was obtained by EA, BCO, GVNS in each of 10 runs. Regarding the execution time, all metaheuristics outperformed the exact solver on average, VND and GVNS being faster for all test instances. Among EA, BCO, VND, and GVNS, the smallest average time over all real-life examples is obtained for the newly proposed GVNS. In order to emphasize the shortest average CPU times, in Table 1.3, these values are presented in bold.

Table 1.4 contains the comparison of results obtained by considered metaheuristic methods on randomly generated test instances. As it was mentioned above, these instances remained out of reach for exact solver from [21]. The first column of Table 1.4 (column heading i) contains the index of the considered instance, while the second column (named *BK*) refers to the best-known cost value. The next four columns contain results related to the EA metaheuristic. In the column named *Best* the best found total cost (obtained after 10 EA executions) is presented. The corresponding average total cost $AvgC$ and average minimum CPU time $AvgT$ are presented in the next two columns.

In order to measure the quality of the obtained EA results, in column $G\%$ we present the average gap calculated as $100 \cdot \frac{AvgC-BK}{BK}$. The results for BCO and GVNS in Table 1.4 are given in the same way. Due to deterministic nature of VND, the column $AvgC$ is omitted, while column T shows the VND running time (in seconds) obtained in a single run. In addition, the gap for VND is calculated as $100 \cdot \frac{Best-BK}{BK}$. In order to highlight the best performing method with respect to the solution quality, the best-known solutions for each instance are bolded in Table 1.4. Similarly, for the best performing method with respect to CPU time, the shortest (average) CPU times for each instance are bolded in Table 1.4.

Table 1.4 Computational results—generated test examples: $l = 35$, $m = 8$, $T = 112$

i	BK	EA				BCO				VND			GVNS			
		$Best$	$AvgC$	$AvgT$	$G\%$	$Best$	$AvgC$	$AvgT$	$G\%$	$Best$	T	$G\%$	$Best$	$AvgC$	$AvgT$	$G\%$
1	**717**	**717**	717.0	104.44	0.0000	718	718.0	143.75	0.1395	**717**	21.16	0.0000	**717**	717.0	**1.04**	0.0000
2	**491**	**491**	491.3	369.58	0.0611	**491**	491.0	54.00	0.0000	493	**1.91**	0.4073	**491**	491.7	49.61	0.1426
3	**683**	**683**	683.6	280.23	0.0878	**683**	683.0	51.80	0.0000	**683**	22.47	0.0000	**683**	683.0	**1.09**	0.0000
4	**554**	**554**	554.0	148.58	0.0000	**554**	554.0	237.64	0.0000	**554**	162.91	0.0000	**554**	555.8	**42.95**	0.3249
5	**594**	**594**	594.0	63.59	0.0000	**594**	594.0	40.41	0.0000	**594**	121.67	0.0000	**594**	594.0	**6.48**	0.0000
6	**486**	**486**	486.0	115.40	0.0000	**486**	486.0	41.42	0.0000	492	**2.38**	1.2346	**486**	486.0	201.44	0.0000
7	**543**	**543**	543.0	133.57	0.0000	**543**	543.0	34.09	0.0000	**543**	203.13	0.0000	**543**	543.0	**21.81**	0.0000
8	**554**	**554**	554.0	351.20	0.0000	**554**	554.0	52.30	0.0000	**554**	3.02	0.0000	**554**	554.0	**0.19**	0.0000
9	**531**	**531**	532.4	364.74	0.2637	**531**	531.0	34.84	0.0000	537	**1.86**	1.1299	**531**	532.2	15.73	0.2260
10	**486**	**486**	486.3	304.19	0.0617	**486**	486.0	42.31	0.0000	**486**	2.00	0.0000	**486**	486.0	**0.14**	0.0000
11	**480**	**480**	480.3	82.47	0.0625	**480**	480.0	190.81	0.0000	**480**	1.92	0.0000	**480**	480.0	**0.14**	0.0000
12	**573**	**573**	573.3	272.73	0.0524	**573**	573.0	145.77	0.0000	578	**2.73**	0.8726	**573**	575.3	327.26	0.4014
13	**520**	**520**	520.0	115.42	0.0000	**520**	520.0	47.23	0.0000	**520**	**9.52**	0.0000	**520**	520.0	94.45	0.0000
14	**557**	**557**	557.0	116.96	0.0000	**557**	557.0	59.77	0.0000	569	**5.02**	2.1544	**557**	560.6	179.80	0.6463
15	**627**	**627**	631.8	266.14	0.7656	**627**	627.0	124.95	0.0000	**627**	6.47	0.0000	**627**	627.0	**0.37**	0.0000
16	**479**	**479**	479.0	125.14	0.0000	**479**	479.0	26.28	0.0000	**479**	385.70	0.0000	**479**	479.0	**8.49**	0.0000
17	**452**	**452**	452.0	135.09	0.0000	**452**	452.0	20.62	0.0000	**452**	71.25	0.0000	**452**	452.0	**9.77**	0.0000
18	**595**	**595**	595.0	134.57	0.0000	**595**	595.0	43.64	0.0000	**595**	6.22	0.0000	**595**	595.0	**0.34**	0.0000
19	**580**	**580**	580.1	159.45	0.0172	**580**	580.0	**53.67**	0.0000	582	104.80	0.3448	**580**	584.6	349.92	0.7931
20	**577**	**577**	577.0	309.37	0.0000	**577**	577.0	150.28	0.0000	583	149.70	1.0399	**577**	581.8	**38.16**	0.8319
21	**623**	**623**	623.0	131.14	0.0000	**623**	623.0	34.84	0.0000	**623**	78.16	0.0000	**623**	623.0	**34.33**	0.0000
22	**495**	**495**	495.3	354.55	0.0606	**495**	495.0	110.23	0.0000	496	**4.28**	0.2020	**495**	495.8	20.87	0.1616
23	**459**	**459**	459.0	138.90	0.0000	**459**	459.0	108.55	0.0000	**459**	225.39	0.0000	**459**	459.0	**23.85**	0.0000
24	**514**	**514**	514.0	139.03	0.0000	**514**	514.0	40.03	0.0000	**514**	121.97	0.0000	**514**	514.0	**21.70**	0.0000

(continued)

Table 1.4 (continued)

i	BK	EA				BCO				VND			GVNS			
		$Best$	$AvgC$	$AvgT$	$G\%$	$Best$	$AvgC$	$AvgT$	$G\%$	$Best$	T	$G\%$	$Best$	$AvgC$	$AvgT$	$G\%$
25	**613**	**613**	613.0	134.45	0.0000	**613**	613.0	**104.45**	0.0000	645	512.64	5.2202	**613**	628.2	190.54	2.4796
26	**477**	**477**	477.0	148.93	0.0000	**477**	477.0	**35.94**	0.0000	**477**	108.45	0.0000	**477**	477.0	40.15	0.0000
27	**517**	**517**	517.0	144.85	0.0000	**517**	517.0	48.69	0.0000	**517**	2.84	0.0000	**517**	517.0	**0.18**	0.0000
28	**517**	**517**	517.0	209.62	0.0000	**517**	517.0	27.19	0.0000	634	560.95	22.6306	**517**	517.0	**21.05**	0.0000
29	**464**	**464**	464.0	132.01	0.0000	**464**	464.0	55.52	0.0000	467	**2.25**	0.6466	**464**	464.3	147.64	0.0647
30	**592**	**592**	592.0	330.95	0.0000	**592**	592.0	32.02	0.0000	**592**	1.81	0.0000	**592**	592.0	**0.13**	0.0000
31	**665**	**665**	665.2	229.23	0.0301	**665**	665.0	200.38	0.0000	675	196.95	1.5038	**665**	672.0	**178.60**	1.0526
32	**495**	**495**	495.0	183.15	0.0000	**495**	495.0	90.05	0.0000	**495**	172.83	0.0000	**495**	495.0	**66.54**	0.0000
33	**481**	**481**	481.0	153.04	0.0000	**481**	481.0	48.23	0.0000	**481**	**6.94**	0.0000	**481**	481.0	14.78	0.0000
34	**539**	**539**	539.0	231.52	0.0000	**539**	539.0	58.23	0.0000	**539**	160.14	0.0000	**539**	539.0	**7.02**	0.0000
35	**528**	**528**	528.0	136.32	0.0000	**528**	528.0	43.98	0.0000	**528**	3.50	0.0000	**528**	528.0	**0.21**	0.0000
36	**522**	**522**	522.6	253.14	0.1149	**522**	522.0	199.94	0.0000	**522**	55.52	0.0000	**522**	522.0	**15.94**	0.0000
37	**467**	**467**	467.0	318.14	0.0000	**467**	467.0	342.61	0.0000	**467**	1.83	0.0000	**467**	467.0	**0.13**	0.0000
38	**479**	**479**	480.4	254.55	0.2923	**479**	479.0	63.61	0.0000	518	**6.66**	8.1420	**479**	483.5	109.43	0.9395
39	**534**	**534**	534.0	136.48	0.0000	**534**	534.0	136.67	0.0000	**534**	40.56	0.0000	**534**	534.0	**2.99**	0.0000
40	**574**	**574**	575.5	114.58	0.2613	**574**	574.0	213.41	0.0000	**574**	**1.25**	0.0000	**574**	574.0	58.36	0.0000
41	**554**	**554**	555.5	477.55	0.2708	**554**	554.0	**60.45**	0.0000	563	237.05	1.6245	**554**	560.7	377.57	1.2094
42	**448**	**448**	448.0	142.48	0.0000	**448**	448.0	35.55	0.0000	**448**	1.77	0.0000	**448**	448.0	**0.13**	0.0000
43	**529**	**529**	534.0	590.94	0.9452	**529**	529.0	**35.47**	0.0000	**529**	151.89	0.0000	**529**	532.9	43.75	0.7372
44	**553**	**553**	553.0	209.99	0.0000	**553**	553.0	62.02	0.0000	**553**	**34.20**	0.0000	**553**	554.2	108.80	0.2170
45	**478**	**478**	478.0	134.73	0.0000	**478**	478.0	72.12	0.0000	**478**	1.86	0.0000	**478**	478.0	**0.13**	0.0000
46	**575**	**575**	575.0	230.33	0.0000	**575**	575.0	109.73	0.0000	**575**	119.75	0.0000	**575**	575.1	**94.92**	0.0174
47	**554**	**554**	558.2	386.90	0.7581	**554**	554.0	**90.06**	0.0000	**554**	256.53	0.0000	**554**	563.3	226.79	1.6787
Average	**538.8**	**538.8**	539.3	212.77	0.0873	538.9	538.9	86.29	0.0030	544.1	92.63	1.0033	**538.8**	540.3	**67.14**	0.2537

As can be seen from Table 1.4, EA and GVNS were able to obtain the best-known solutions for all test instances. BCO missed *BK* solution only for the first instance, while VND failed to reach the best-known solution for 14 out of 47 instances. In addition, VND produced the largest gap with respect to the best-known solution. With an exception of the first instance, BCO produced the *BK* solution in all ten runs, as it can be seen from the corresponding $G\%$ column. EA and GVNS have small values for average gap, with EA being slightly more reliable method. The same conclusion holds with respect to the average cost values. However, BCO still performs better with respect to both $AvgC$ and $G\%$.

Regarding the (average) minimum CPU time, the superior method is GVNS, followed by BCO, VND, and EA. The (average) minimum CPU times of GVNS, BCO, VND, and EA are 67.14, 86.29, 92.63, and 212.77 s, respectively. This means that GVNS is 28.5% faster than BCO, 37.97% faster than VND, and 216.9% faster than EA.

From the presented computational results, it can be seen that GVNS outperforms other methods on both data sets regarding the execution speed. Having in mind that the average gap values are quite small for all methods, we can consider GVNS to be the most suitable method for MCHBAP.

1.7 Conclusion

We studied Minimum Cost Hybrid Berth Allocation Problem (MCHBAP), a variant of the well known BAP that is of great practical importance in maritime transportation. The design of efficient and reliable decision support system in container terminal heavily depends on the speed of finding high-quality solutions for underlying BAP. In real-life applications, MCHBAP is highly dynamic and it is important to provide high quality solution in short CPU time. Having in mind that this problem is NP hard, it should be addressed by metaheuristic methods. Therefore, we presented three metaheuristic approaches: Evolutionary Algorithm [24], Bee Colony Optimization [22], Variable Neighborhood Descent [7], and proposed a General Variable Neighborhood Search for MCHBAP. In order to compare the efficiency of the four metaheuristic methods (EA, BCO, VND, GVNS), against each other and against exact approach from [21], two sets of test instances were considered: a set of real-life instances and randomly generated ones. As the metaheuristics are generally stochastic methods (except VND in our case), we examined their stability by performing multiple runs.

The experimental results obtained on real-life instances show that in each run all four metaheuristics were able to find optimal solutions in shorter running times than exact solver. The newly proposed GVNS method showed to be significantly faster compared to BCO, EA, and VND, while VND was faster than BCO and EA in reaching optimal solutions. Randomly generated test examples showed to be extremely hard, because exact solver [21] was not able to find their optimal solutions. On the other hand, all considered metaheuristic approaches easily provided high-quality solutions for all generated instances in each run. Experimental results on this data set indicate that the proposed GVNS outperforms other three metaheuristic methods regarding CPU time, while solution quality remains at high level.

Our computational results indicate that GVNS has obvious advantage over presented methods for MCHBAP. We strongly believe that it may also be applied to similar problems in maritime transportation. For further enhancing of metaheuristics, their hybridization and combination with exact methods seem to be encouraging topics.

Acknowledgements This research was partially supported by Serbian Ministry of Education, Science and Technological Development under the grants nos. 174010 and 174033. The authors would like to thank the anonymous referees for the valuable suggestions that led to the improved presentation of the results described in this manuscript.

References

1. Bierwirth, C., Meisel, F..: A fast heuristic for quay crane scheduling with interference constraints. J. Sched. **12**(4), 345–360 (2009)
2. Bierwirth, C., Meisel, F.: A survey of berth allocation and quay crane scheduling problems in container terminals. Eur. J. Oper. Res. **202**, 615–627 (2010)
3. Bierwirth, C., Meisel, F.: A follow-up survey of berth allocation and quay crane scheduling problems in container terminals. Eur. J. Oper. Res. **244**(3), 675–689 (2015)
4. Chang, D., Jiang, Z., Yan, W., He, J.: Integrating berth allocation and quay crane assignments. Transp. Res. Part E **46**(6), 975–990 (2010)
5. Cheong, C.Y., Tan, K.C.: A multi-objective multi-colony ant algorithm for solving the berth allocation problem. In: Advances of Computational Intelligence in Industrial Systems, pp. 333–350. Springer (2008)
6. Cordeau, J.F., Laporte, G., Legato, P., Moccia, L.: Models and tabu search heuristics for the berth-allocation problem. Transp. Sci. **39**(4), 526–538 (2005)
7. Davidović, T., Kovač, N., Stanimirović, Z.: VNS-based approach to minimum cost hybrid berth allocation problem. In: Proceedings of SYMOPIS 2015, pp. 237–240, Silver Lake, Serbia (2015)
8. Davidović, T., Lazić, J., Mladenović, N., Kordić, S., Kovač, N., Dragović, B.: Mip-heuristics for minimum cost berth allocation problem. In: Proceedings of International Conference on Traffic and Transport Engineering, ICTTE 2012, pp. 21–28, Belgrade, Serbia (2012)
9. Davidović, T., Teodorović, D., Šelmić, M.: Bee colony optimization part I: the algorithm overview. Yugoslav J. Oper. Res. **25**(1), 33–56 (2015)
10. De Falco, I., Della Cioppa, A., Tarantino, E.: Mutation-based genetic algorithm: performance evaluation. Appl. Soft Comput. **1**(4), 285–299 (2002)
11. Filipović, V.: Fine-grained tournament selection operator in genetic algorithms. Comput. Artif. Intell. **22**(2), 143–161 (2003)
12. Goh, K.-S., Lim, A.: Combining various algorithms to solve the ship berthing problem. In: 12th IEEE International Conference on Tools with Artificial Intelligence, ICTAI 2000, pp. 370–375 (2000)
13. Guan, Y., Cheung, R.K.: The berth allocation problem: models and solution methods. OR Spectr. **26**(1), 75–92 (2004)
14. Han, M., Li, P., Sun, J.: The algorithm for berth scheduling problem by the hybrid optimization strategy gasa. In: 9th International Conference on Control, Automation, Robotics and Vision, pp. ICARCV'06., 1–4. IEEE (2006)

15. Hansen, P., Mladenović, N., Brimberg, J., Moreno Pérez, J.A.: Variable neighbourhood search. In: Gendreau, M., Potvin, J.-Y. (eds.) Handbook of Metaheuristics, 2nd edn., pp. 61–86. Springer, New York (2010)
16. Hansen, P., Oğuz, C., Mladenović, N.: Variable neighborhood search for minimum cost berth allocation. Eur. J. Oper. Res. **191**(3), 636–649 (2008)
17. Holland, J.H.: Adaptation in Natural and Artificial Systems. The University of Michigan Press, Ann Arbor (1975)
18. Imai, A., Nishimura, E., Hattori, M., Papadimitriou, S.: Berth allocation at indented berths for mega-containerships. Eur. J. Oper. Res. **179**(2), 579–593 (2007)
19. Imai, A., Nishimura, E., Papadimitriou, S.: The dynamic berth allocation problem for a container port. Transp. Res. Part B **35**, 401–417 (2001)
20. Kim, K.H., Moon, K.C.: Berth scheduling by simulated annealing. Transp. Res. Part B **37**(6), 541–560 (2003)
21. Kordić, S., Dragović, B., Davidović, T., Kovač, N.: A combinatorial algorithm for berth allocation problem in container port. In: The 2012 International Association of Maritime Economists Conference, IAME 2012, Taipei (2012)
22. Kovač, N.: Bee colony optimization algorithm for the minimum cost berth allocation problem. In: XI Balcan Conference on Operational Research, pp. 245–254, BALCOR 2013, Beograd-Zlatibor, Serbia (2013)
23. Kovač, N.: Metaheuristic approaches for the berth allocation problem. Yugoslav J. Oper. Res. (2017). doi:10.2298/YJOR160518001K
24. Kovač, N., Davidović, T., Stanimirović, Z.: Evolutionary algorithm for the minimum cost hybrid berth allocation problem. In: IEEE Conference publications IISA 2015, pp. 1–6, Ionian University, Corfu, Greece (2015)
25. Lim, A.: The berth planning problem. Oper. Res. Lett. **22**(2), 105–110 (1998)
26. Marić, M., Stanimirović, Z., Stanojević, P.: An efficient memetic algorithm for the uncapacitated single allocation hub location problem. Soft Comput. **17**(3), 445–466 (2013)
27. Meisel, F.: Seaside Operations Planning in Container Terminals. Springer, Berlin (2009)
28. Mladenović, N., Hansen, P.: Variable neighborhood search. Comput. Oper. Res. **24**(11), 1097–1100 (1997)
29. Moorthy, R., Teo, C.-P.: Berth management in container terminal: the template design problem. OR Spectr. **28**(4), 495–518 (2006)
30. Murata, H., Fujiyoshi, K., Nakatake, S., Kajitani, Y.: VLSI module placement based on rectangle-packing by the sequence-pair. IEEE Trans. Comput.-Aided Des. Integr. Circ. Syst. **15**(12), 1518–1524 (1996)
31. Nishimura, E., Imai, A., Papadimitriou, S.: Berth allocation planning in the public berth system by genetic algorithms. Eur. J. Oper. Res. **131**, 282–292 (2001)
32. Oğuz, C., Narin, Ö.: Solving berth allocation problem with column generation. In: Multidisciplinary International Conference on Scheduling : Theory and Applications, MISTA 2009, pp. 744–747, Dublin, Ireland (2009)
33. Park, Y.M., Kim, K.H.: A scheduling method for berth and quay cranes. OR Spectr. **25**(1), 1–23 (2003)
34. Prodhon, C.: A hybrid evolutionary algorithm for the periodic location-routing problem. Eur. J. Oper. Res. **210**, 204–212 (2011)
35. Rashidi, H., Tsang, E.P.K.: Novel constraints satisfaction models for optimization problems in container terminals. Appl. Math. Model. **37**(6), 3601–3634 (2013)
36. Stahlbock, R., Voß, S.: Operations research at container terminals: a literature update. OR Spectr. **30**(1), 1–52 (2008)
37. Stanimirović, Z., Marić, M., Božović, S., Stanojević, P.: An efficient evolutionary algorithm for locating long-term care facilities. Inf. Technol. Control **41**(1), 77–89 (2012)
38. Tang, X., Tian, R., Wong, D.F.: Fast evaluation of sequence pair in block placement by longest common subsequence computation. IEEE Trans. Comput.-Aided Des. Integr. Circ. Syst. **20**(12), 1406–1413 (2001)

39. Teodorović, D., Šelmić, M., Davidović, T.: Bee colony optimization part II: the application survey. Yugoslav J. Oper. Res. **25**(2), 185–219 (2015)
40. Theofanis, S., Boile, M., Golias, M.: An optimization based genetic algorithm heuristic for the berth allocation problem. IEEE Congr. Evol. Comput. DCEC **2007**, 4439–4445 (2007)
41. Ting, C.-J., Wu, K.-C., Chou, H.: Particle swarm optimization algorithm for the berth allocation problem. Expert Syst. Appl. **41**(4), 1543–1550 (2014)
42. Wang, F., Lim, A.: A stochastic beam search for the berth allocation problem. Decis. Support Syst. **42**, 2186–2196 (2007)

Chapter 2
Ship Stability Considerations in the Quay Crane Scheduling Problem

Noura Al-Dhaheri

Abstract The aim of this paper is to present the recent literature that has advanced in the field of maritime logistics, specifically with regards to the consideration of vessel stability during the process of unloading and/or loading containers onto vessels. This process is essentially known as the Quay Crane Scheduling Problem (QCSP) which determines the operational profile of each quay crane in terms of the container tasks and timing. The literature on this and other problems pertaining to quayside operational planning is presented, before introducing the works with a contribution in ship stability. The works are described with insights into their formulation and solution techniques, as well as their contribution to the literature. Most importantly, the results are discussed and directions are provided for future work in the area.

2.1 Research in Maritime Logistics

The field of maritime logistics involves a number of important operational problems which have attracted the interest of researchers in recent years, due to continuously increasing volumes of containerized cargo. Attention is largely focused on quayside operations, which refer to the sequence of activities taking place from the arrival of a ship within the seaside of the port to the handling of its container load to the yard-side. High costs of berthing and container handling incurred on vessel operators, as well as the high costs of operating container terminal equipment create the need for efficient processes, able to accommodate the high demand at minimum cost [25].

Container terminals comprise four main areas, namely the quay, the buffer, the yard and the gate. An overview of the container terminal layout is presented in Fig. 2.1. The quay and the buffer areas are considered seaside, while the yard and gate areas are considered landside as can be found in [7]. Once a vessel is berthed, a certain number of quay cranes (QC) are employed to discharge import and transshipment containers from the vessel to the quay and/or load export and transshipment

N. Al-Dhaheri (✉)
Maqta Gateway, Abu Dhabi Ports, Abu Dhabi, United Arab Emirates
e-mail: noura.aldhaheri@adports.ae

C. Konstantopoulos and G. Pantziou (eds.), *Modeling, Computing and Data Handling Methodologies for Maritime Transportation*, Intelligent Systems Reference Library 131, DOI 10.1007/978-3-319-61801-2_2

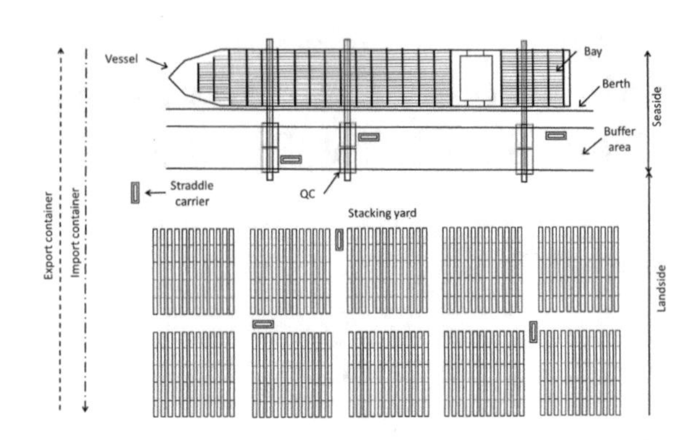

Fig. 2.1 Overview of container terminal layout

containers from the quay onto the vessel. The vessel is longitudinally divided into several bays and, each container is discharged from or loaded onto a given bay according to a stowage plan. Import, transshipment and export containers are temporarily stored in the container stacking yard. Import containers are transported from the quay to the stacking yard where they stay till being delivered to customers through the gate. Transshipment and export containers are stored in the stacking yard until the vessel that ships them to their next destination arrive. They are subsequently transported to the quay. Container transport from the quay to the stacking yard and vice versa is ensured by shuttle vehicles such as internal trucks, automated guided vehicles and straddle carriers [2]. For a more comprehensive overview on the layout of container terminals and, the handling and transport equipment, the reader can refer to [23] and **brinkman 2011**. Quayside problems include the Berth Allocation Problem (BAP) which determines the berths that incoming vessels are assigned to, the Quay Crane Assignment Problem (QCAP), whereby the required cranes are assigned to each ship, before scheduling of crane tasks can take place through the Quay Crane Scheduling Problem (QCSP).

The BAP is one of the well-studied problems in the literature of maritime logistics. It can be distinguished into the static BAP and dynamic BAP, depending on the assumed arrival profile of ships. The static case of the BAP is addressed by Simrin et al. [21], who develop a Lagrangian relaxation based approach for the problem, while the dynamic case is tackled by Arango et al. [7] and Schoonenberg et al. [19]. Other authors focus on service priority agreements between port and vessel operators [6, 11], and explicitly account for these in their formulations, as these are often very expensive agreements. Aside from containers terminals, the BAP has also been studied in the context of bulk ports [1], in which additional challenges arise with respect to the decision of a ship's berthing position, depending on the facilities and equipment required for the handling of various types of cargo.

The QCAP is frequently integrated with either the BAP or the QCSP, due to the strong interdependence with each of the aforementioned problems. An example of the integration of the BAP with the QCAP is the work of [15]. The QCSP is known to be the most complex among the three quayside operational problems, due to the fact that tasks must be scheduled while strict physical constraints are satisfied. Specifically, given that cranes travel on a single rail, non-crossing must be enforced at all times. In addition, cranes must be positioned at least certain bays apart to prevent interference. Therefore, the QCSP is inherently complex, leading the majority of researchers to develop heuristic techniques to tackle it.

The first notable work on the QCSP is that of [13], who develop a branch-and-bound and a Greedy Randomized Adaptive Search Procedure (GRASP) to solve the problem with the objective of minimizing the weighted sum of the makespan and QCs completion times. Their model was later refined by Moccia et al. [17] who account for safety margin constraints in a more stringent way and solve medium and large instances of the problem using branch-and-cut (B&C). Other heuristics used include Tabu Search (TS) [18], tree-search-based heuristics [16] and Genetic Algorithm (GA) [14], among others. As far as exact techniques are concerned, [12] develop a Branch-and-Price algorithm to solve the problem to optimality.

Due to the strong interdependency between all three problems pertaining to quay-side operations, and especially between the assignment and scheduling problem, researchers have investigated the integration of these two problems, rather than treating them only sequentially. The resulting problem is known as the integrated Quay Crane Assignment and Scheduling Problem (QCASP), for which researchers have presented novel approaches both in terms of problem formulation and solution techniques that aim to effectively and efficiently solve this inherently complex problem. Integration consistently leads to better results but it comes at the cost of increased computational intensity, which is an issue that many researchers attempt to address. Diabat and Theodorou [8, 22] try to overcome the complexity of the integrated problem by transforming crane scheduling into a crane-to-bay allocation problem and solve the problem using heuristics, namely a Genetic Algorithm (GA) and Lagrangian-relaxation based heuristic, respectively. The authors assume a uni-directional crane movement to model the problem. Bi-directional movement is adopted by Fu et al. [10] and Fu and Diabat [9] who also solve the integrated QCASP using a GA and Lagrangian relaxation-based heuristic, respectively. Other researchers, such as [19] approach the integrated QCASP from a cost perspective, aiming to indicate the impact of load time cost on vessel's prioritization. The authors allow for adjustable QC operational costs as well as adjustable working rates, as quay cranes are not necessarily identical.

The continuous advancement of research gives rise to more practical operational aspects which were not previously considered. One such aspect is the issue of vessel stability during the process of handling a container vessel, which is presented in more detail in the following section, both in terms of problem description and literature contribution.

2.2 Ship Stability Consideration

To provide a visual illustration of the significance of ship stability, Fig. 2.2 depicts the great imbalance that can be caused in the case of uneven container weight distribution across the vessel. The stability of the vessel is violated if the vessel's center of

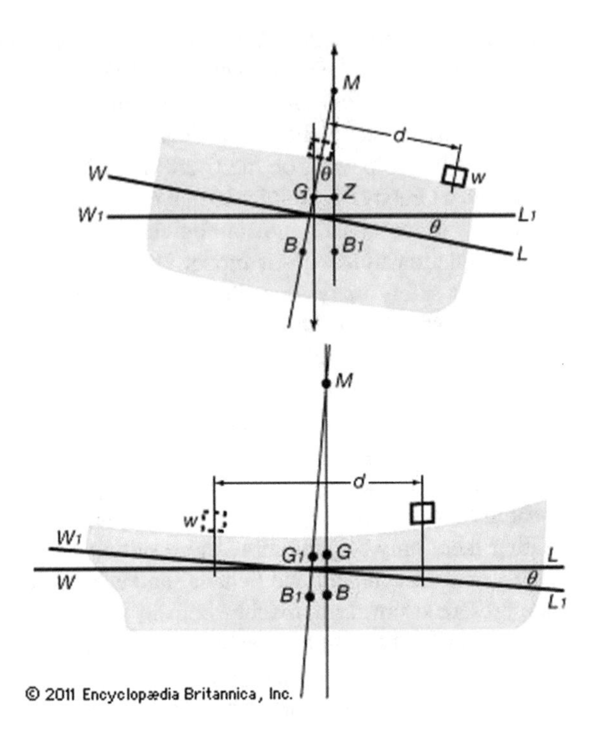

Fig. 2.2 Illustration of imbalance during container loading/unloading process

gravity shifts too much toward one side during the loading or unloading process. This shift results from workload distribution along the vessel and the sequence of operations. Henceforth the QCSP should consider vessel stability constraints to allow for obtaining QC schedules that can be used in practice.

In 2013, [24] produced the first paper that highlighted the importance of considering vessel stability in the QCSP. Wang et al. [24] incorporate vessel stability into a mixed integer program (MIP) and develop a Genetic Algorithm (GA) to solve the problem. The GA discards any QC schedule that violates the QCSP constraints, including those related to vessel stability.

Al-Dhaheri and Diabat [2] develop a Mixed Integer Program (MIP), in which they implicitly account for vessel stability by minimizing the differences between the container loads stacked over a number of bays and by maintaining a balanced load across all bays. Furthermore, this MIP accounts for important considerations such as the bidirectional movement of cranes and the ability to move between bays even before completion of all container tasks. The increased complexity of these additional considerations is offset by the simple objective, and overall the model is proved to successfully address the vessel stability issue and it is solvable for small- to medium-sized instances. While typically in the literature the objective of the QCSP is to minimize the "makespan" of the QC schedule, the contribution of the aforementioned paper is that it proves that this objective is equally served by ensuring sufficient balance of the workload distribution among the cranes.

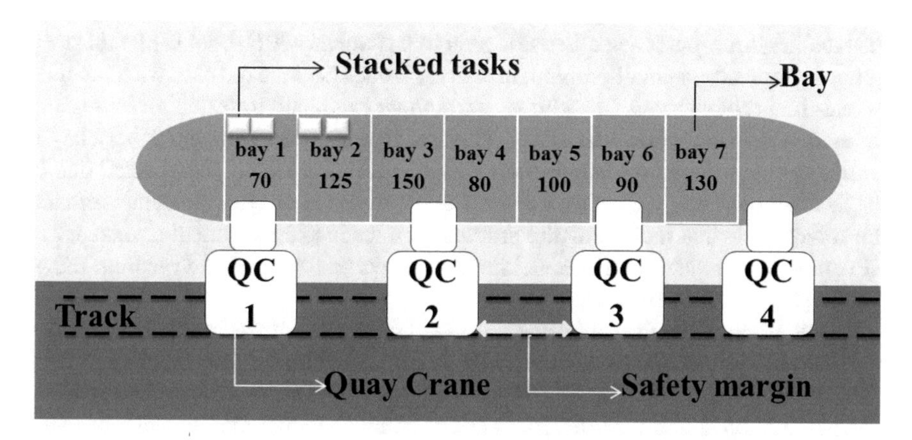

Fig. 2.3 Graphical illustration of the container weight distribution during the QCSP

Unlike the majority of models developed in the literature, the current model allows for QCs to travel between bays, even before the container tasks at that bay are completed. In fact, this is readily implemented in practice because this extra degree of freedom allows for better solutions. This leads to a model that better reflects real-life circumstances, as long as the non-crossing constraints are maintained at all times. Given the fact that the time required for cranes to travel between bays is negligible, compared to the handling times, it is not taken into consideration. However, the authors illustrate how this can be easily incorporated as an extension to the model. The authors also relax the restriction of unidirectional crane movement that is frequently adopted by researchers. Thus, the authors allow for fully taking advantage of the assumption that cranes can move between bays, even before the completion of all container handling at the currently assigned bay. Finally, identical service rates are assumed for all cranes, but once again the authors demonstrate that the formulation can be modified to account for variability in service rates.

Figure 2.3 provides a graphical illustration of the problem, depicting the sequential indexing along the quay for both bays and cranes, the rail upon which cranes are mounted, the safety margins between adjacent cranes, as well as the stacked containers on the first bay. When dividing the vessels into bay areas, the safety margin between two adjacent QCs can be implicitly taken into consideration; hence the safety margin has not been incorporated into this model.

The authors conduct a computational analysis through which several important insights are drawn: first of all, a lower number of cranes with higher efficiency is a preferred option compared to a higher number of cranes with lower efficiency. A second conclusion is that changing the distribution of container workload across bays has an impact on the total handling time required to serve the vessel; in fact, the greater the standard deviation of the container workload difference between bays, the greater the handling time required. Another interesting observation made by the authors' analysis is that reordering of container workload does not have an impact on the handling time, but it actually has a great impact on the CPU time. While this may not be significant for small problem instances, it does become the case when applied to real-life problem sizes. In addition, the authors examined the effect of increasing the number of bays, while maintaining the same workload and concluded that a larger number of bays do not lead to a lower handling time, as would be expected due to the fact that more bays allow for more flexible movement of QCs. As a final remark, the authors note that increasing the problem size leads to an exponential increase in CPU time, thus rendering the model more appropriate for small and medium rather than large sized problems.

In the work of [4] the authors aim to explicitly rather than implicitly consider vessel stability. Once again, a novel MIP is developed that is very flexible in handling various settings of the QCSP, such as those related to crane traveling time, task preemption, and the unidirectional quay crane operating mode. While the MIP formulation provides an optimal unidirectional schedule for all considered small-sized problem, in relatively reasonable time, it does not return the optimal solution for all these instances. Moreover, the experimentation of the MIP formulation on small-sized problems and the use of the proposed lower bound (LB), highlight the interest for adopting a search strategy focused on unidirectional schedules.

A Genetic Algorithm is also designed to solve the problem. The GA embraces the unidirectional search strategy. It provides optimal unidirectional schedules for all small-sized instances within significantly lower time than the one required by GAMS. Furthermore, for medium and large size instances, where GAMS fails to solve the problem, the GA returns a near-optimal solution within a reasonable computational time. This promotes the use of the proposed GA as a solution approach for the considered QCSP, especially for medium and large sized instances.

One aspect that remains to be addressed is the inherent uncertainty associated with the scheduling problem. This is addressed in the work of [4], who develop a formulation that incorporates the randomness related to the handling rates and idle times of quay cranes and stacking cranes. The objective is to minimize vessel handling time while considering the entire container handling process involving both seaside operations and container transfer operations, tasks that take place between the quay and the stacking yard. A stochastic mixed integer programming model is proposed and a simulation based Genetic Algorithm (GA) is applied to construct QC schedules that account for the dynamics and the uncertainty inherent to the container handling process. The proposed GA framework embeds a simulation model to evaluate the fitness of each chromosome. The proposed algorithm is tested under both stochastic and deterministic circumstances. The obtained solutions are furthermore evaluated more accurately using the simulation model with a larger sample size. Simulation results show that the algorithm provides better QC schedules when it is used under stochastic environment. However, the algorithm incurs much larger computational time when it is used under stochastic environment than when it is used under deterministic environment.

The authors point out that the intended QC performance in terms of utilization rate cannot be reached without employing a sufficient number of Straddle Carriers (SCs). Moreover, the results highlight the significance of using a simulation model to obtain more realistic and reliable performance of the QC schedules returned by the algorithm under deterministic environment. Computational experiments demonstrate satisfactory results of the proposed algorithm and stress the importance of simulation in obtaining more reliable estimates of QC schedule performance.

The final model presented in the current review is that of [3], which encompasses three major contributions: first, to develop a new and more tractable optimization model for the single-vessel QCSP with vessel stability constraints; secondly, to extend this model to the multiple-vessel QCSP; and thirdly, to design an efficient solution algorithm that is capable of solving real-sized instances of the problem. In order to achieve these research milestones, the QCSP is initially formulated for the single vessel case, with the objective of minimizing the makespan without considering stability constraints. After devising a new formulation for the single vessel case (whose performance is evaluated against benchmark formulations of the literature), the authors use its solution in another optimization problem that considers the stability constraints. Then they extend the model to the multiple vessel case, which is solved with the help of Lagrangian relaxation, whereby the problem is decomposed by vessel and each is solved efficiently as a single vessel case. The Lagrangian subproblems don't have the integrity property, and therefore the solution of the Lagrangian subproblems provides a lower bound, that is at least as good as the linear programming relaxation bound, on the optimal value of the original problem. The Lagrangian multipliers are updated using the cutting plane method and the solution of the Lagrangian master problem provides an upper bound on the optimal value of the Lagrangian lower bound. Upper bounds on the optimal value of the original problem objective function are obtained using a constructive heuristic, and through computational experiments we demonstrate the performance of the Lagrangian relaxation-based procedures.

First the authors present the single ship case: for a given ship, and a number of identical quay cranes (QCs) Q, the aim of the QCSP with ship stability considerations is to schedule the work of cranes in a way that all tasks are performed. A task is defined on the basis of the unloading or loading operation of a single container. Preemption is permitted, as a single bay can be assigned to multiple QCs for the handling of its containers. However, at any time, there can be at most one QC working on any bay. This is enforced through what are known as interference constraints. In addition, all QCs travel on a single rail along the quay, which implies that crossing is not allowed. We assume that the traveling time between two consecutive bays is constant, and identical for all QCs. A safety margin is also maintained which is measured by the minimum distance in bays between adjacent QCs.

The authors extend the formulation from the single ship to the multi ship case, assuming in this case that the berthing order is known for the ships to be handled, i.e. the static case of the problem is addressed. In addition, they assume that cranes can move between ships, which means it is not necessary for a crane to complete all operations at the current ship before moving on to the next.

2.3 Outlook and Future Work

As far as future research is concerned, there is great potential for expansion of the models presented addressing the issue of ship stability. There are several assumptions that can be easily included, such as a non-constant productivity rate for cranes and the consideration of the travel times required for cranes to travel between bays. In addition, non-identical container weights could be assumed, as is the case in practice, which would pose additional modeling challenges in terms of weight distribution. Another modeling contribution would be to assume both unloading and loading operations.

This work can be further enhanced by integrating the QCSP with stability constraints with the Berth Allocation Problem (BAP). Until now this has been a challenge in the field, due to the very high complexity of the integrated problem. However, extending the current formulation and implementing a similar solution approach could lead to an efficient solution of the problem. A future direction for the solution approach could be to apply Branch & Price (B&P) to solve the single ship problem with stability constraints in one stage rather than in a two stage approach involving a heuristic. Then, the multi-ship problem would be solved once again with Lagrangian relaxation and with the subproblems solved using B&P.

For a truly comprehensive formulation, other aspects of container terminal operations could be accounted for, such as yard congestion. This occurs when the yard storage areas are overly accessed by vehicles transporting containers to respective stacks. An interesting extension would be to consider lateral in addition to horizontal stability, which would require a three dimensional approach to the problem. Finally, the QCSP could be combined with the Container Relocation Problem (CRP), which involves placing container on other bays while containers deeper in the stack of the current bay are being accessed. Also, given that the QCs represent one of the major valuable resources at container terminals, it is worth further investigating the SCs to assign to each QC while taking into account more details regarding yard congestion and containers' stack location. Also, it could be useful to investigate other SCs deployment strategies such as pooling, where SCs are assigned to vessels rather than to QCs. Thus, within a pooling strategy, any SC can serve any QC assigned to the vessel.

As far as solution techniques are concerned, more heuristics can be developed and compared to the existing ones, in terms of performance and efficiency. Finally, conducting tests on real-life instances that have been used by other works and benchmarking them would be extremely beneficial in terms of comparatively evaluating the model and ultimately judging its appropriateness for use by container terminal operators.

References

1. Al Hammadi, J., Diabat, A.: An integrated berth allocation and yard assignment problem for bulk ports: formulation and case study. RAIRO Oper. Res. (2017) (in press)
2. Al-Dhaheri, N., Diabat, A.: The quay crane scheduling problem. J. Manuf. Syst. **36**, 87–94 (2015)
3. Al-Dhaheri, N., Diabat, A.: A Lagrangian-relaxation-based heuristic for the multiship quay crane scheduling problem with ship stability constraints. Ann. Oper. Res. (2016)
4. Al-Dhaheri, N., Jebali, A., Diabat, A.: A simulation based Genetic Algorithm approach for the quay crane scheduling under uncertainty. Simul. Model. Pract. Theory **66**, 122–138 (2016a)
5. Al-Dhaheri, N., Jebali, A., Diabat, A.: The quay crane scheduling problem with nonzero crane repositioning time and vessel stability constraints. Comput. Ind. Eng. **94**, 230–244 (2016b)
6. Alzaabi, S., Diabat, A.: On the berth allocation problem. RAIRO Oper. Res. **50**(3), 491–501 (2016)
7. Arango, C., Corts, P., Onieva, L., Escudero, A.: Simulation-optimization models for the dynamic berth allocation problem. Comput. Aided Civil Infrastruct. Eng. **28**(10), 769–779 (2013)
8. Diabat, A., Theodorou, E.: An integrated quay crane assignment and scheduling problem. Comput. Ind. Eng. **73**, 115–123 (2014)
9. Fu, YM., Diabat, A.: A Lagrangian relaxation approach for solving the integrated quay crane assignment and scheduling problem. Appl. Math. Model. **39**(3–4), 1194–1201 (2015)
10. Fu, YM., Diabat, A., Tsai, IT.: A multi-vessel quay crane assignment and scheduling problem: formulation and heuristic solution approach. Expert Syst. Appl. **41**(15), 6959–6965 (2014)
11. Imai, A., Nishimura, E., Papadimitriou, S.: Berth allocation with service priority. Transp. Res. Part B: Methodol. **37**(5), 437–457 (2003)
12. Kenan, N., Diabat, A.: A branch-and-price algorithm to solve a quay crane scheduling problem. Proc. Comput. Sci. **61**, 527–532 (2015)
13. Kim, K.H., Park, Y.M.: A crane scheduling method for port container terminals. Eur. J. Oper. Res. **156**(3), 752–768 (2004)
14. Liang, C., Huang, Y., Yang, Y.: A quay crane dynamic scheduling problem by hybrid evolutionary algorithm for berth allocation planning. Comput. Ind. Eng. **56**(3), 1021–1028 (2009)
15. Lu, Z., Han, X., Xi, L.: Simultaneous berth and quay crane allocation problem in container terminal. Adv. Sci. Lett. **4**(6–7), 2113–2118 (2011)
16. Meisel, F.: The quay crane scheduling problem with time windows. Naval Res. Logist. (NRL) **53**(1), 45–59
17. Moccia, L., Cordeau, JF., Gaudioso, M., Laporte, G.: A branch-and-cut algorithm for the quay crane scheduling problem in a container terminal. Naval Res. Logist. **53**(1), 45–59 (2006)
18. Sammarra, M., Cordeau, JF., Laporte, G., Monaco, MF.: A Tabu search heuristic for the quay crane scheduling problem. J. Sched. **10**(4–5), 327–336 (2007)
19. Schoonenberg, W., Hols, J., Diabat, A.: A cost based approach for a crane assignment and scheduling problem. In: International Conference on Industrial Engineering and Systems Management (IESM), 21–23 Oct 2015, Seville, Spain
20. Simrin, A., Diabat, A.: The dynamic berth allocation problem: a linearized formulation. RAIRO Oper. Res. **49**(3), 473–494 (2015)
21. Simrin, A.S., Alkawaleet, N.N., Diabat, A.H.: A Lagrangian relaxation based Heuristic for the static berth allocation problem using the cutting plane method. In: Proceedings of the 15th International Conference on Enterprise Information Systems, pp. 565–569 (2013)
22. Theodorou, E., Diabat, A.: A joint quay crane assignment and scheduling problem: Formulation, solution algorithm and computational results. Optim. Lett. **9**, 799–817 (2015)
23. Vo, S., Stahlbock, R., Steenken, D.: Container terminal operation and operations research—a classification and literature review. OR Spectrum **26**(1), 3–49 (2004)
24. Wang, J., Hu, H., Song, Y.: Optimization of quay crane scheduling constrained by stability of vessels. Transp. Res. Rec.: J. Transp. Res. Board **2330**(1), 47–54 (2013)
25. Zeng, Q., Diabat, A., Zhang, Q.: A simulation optimization approach for solving the dual-cycling problem in container terminals. Marit. Policy Manag. **42**(8), 87–94 (2015)

Chapter 3
Recent Progress Using Matheuristics for Strategic Maritime Inventory Routing

Dimitri J. Papageorgiou, Myun-Seok Cheon, Stuart Harwood, Francisco Trespalacios and George L. Nemhauser

Abstract This chapter presents an extensive computational study of simple, but prominent matheuristics (i.e., heuristics that rely on mathematical programming models) to find high quality ship schedules and inventory policies for a class of maritime inventory routing problems. Our computational experiments are performed on a test bed of the publicly available MIRPLib instances. This class of inventory routing problems has few constraints relative to some operational problems, but is complicated by long planning horizons. We compare several variants of rolling horizon heuristics, K-opt heuristics, local branching, solution polishing, and hybrids thereof. Many of these matheuristics substantially outperform the commercial mixed-integer programming solvers CPLEX 12.6.2 and Gurobi 6.5 in their ability to quickly find high quality solutions. New best known incumbents are found for 26 out of 70 yet-to-be-proved-optimal instances and new best known bounds on 56 instances.

Keywords Deterministic inventory routing · Matheuristics · Maritime transportation · Mixed-integer linear programming · Time decomposition

3.1 Introduction

In 2014, the volume of world seaborne shipments was estimated to be 9.84 billion tons, accounting for roughly 80% of total world merchandise trade [48]. The petrochemical sector was responsible for transporting approximately 32% of this tonnage with 17% going to crude oil, 9% to petroleum products, and 6% to gas and chemicals [48, Fig. 1.3]. The transportation costs associated with these commodities directly

D.J. Papageorgiou (✉) · M.-S. Cheon · S. Harwood · F. Trespalacios
Corporate Strategic Research, ExxonMobil Research and Engineering Company,
1545 Route 22 East, Annandale, NJ 08801, USA
e-mail: dimitri.j.papageorgiou@exxonmobil.com

G.L. Nemhauser
H. Milton Stewart School of Industrial and Systems Engineering,
Georgia Institute of Technology, 765 Ferst Drive NW, Atlanta, GA 30332, USA

C. Konstantopoulos and G. Pantziou (eds.), *Modeling, Computing and Data Handling Methodologies for Maritime Transportation*, Intelligent Systems Reference Library 131, DOI 10.1007/978-3-319-61801-2_3

affect the economic viability of certain projects and thus effective supply chain decision support tools are critical to ensure profitability. This chapter discusses recent progress using matheuristics (i.e., heuristics that rely on a mathematical programming model) to find high quality ship schedules and inventory policies for a certain maritime transportation problem known as the Maritime Inventory Routing Problem (MIRP), which plays a critical part in global bulk shipping. As there have been several papers over the last 5 years that have investigated matheuristics for particular (often proprietary) applications, we review and evaluate several prominent methods on a publicly available data set.

Inventory routing problems (IRPs) comprise the integration and coordination of two components of the logistics value chain: inventory management and vehicle routing. After their emergence in the industrial gases industry, IRPs are now prominent in a number of business sectors with maritime IRPs arising in the petrochemical sector being a primary source of real-world IRP applications [15, p. 2]. IRPs have gained an important status because they are instrumental in vendor managed inventory (VMI), a policy in which a central decision maker oversees both the inventory and its distribution within a supply chain [12]. Andersson et al. [5] furnish a summary of research on IRPs in road and maritime contexts. Christiansen et al. [14] offer a thorough overview of maritime transportation.

Relative to other transportation industries (e.g., air, rail, and truck), the maritime sector has been slow to adopt optimization-based decision support tools for business planning and operations. As a consequence, relatively few commercial software tools have been developed to fill this niche market. Meanwhile, the complexity of such problems is often too great to be handled using spreadsheets and experience.

Matheuristics are heuristic algorithms made by the interoperation of metaheuristics and mathematical programming techniques [10]. They have garnered increasing attention in the last decade due to advances in mathematical programming software that can now reliably solve certain classes of optimization problems to near optimality within an acceptable time limit [9]. Matheuristics for IRPs are discussed in [8] and, more generally, for vehicle routing problems in [6]. They have been applied in other shipping applications, such as in liner shipping network design Brouer et al. [11].

The guiding tenet behind the matheuristics presented here is the desire to maintain a single mixed-integer linear programming (MILP) formulation and to find high quality solutions to it by iteratively fixing subsets of variables and solving the resulting MILP with a general purpose MILP solver. Specifically, consider a generic MILP

$$\min \ \mathbf{c}^\top \mathbf{x} \tag{3.1a}$$

$$\text{s.t.} \ \mathbf{A}\mathbf{x} \geq \mathbf{b} \tag{3.1b}$$

$$\mathbf{x} \in \mathbb{Z}^{n_1} \times \mathbb{R}^{n_2} \tag{3.1c}$$

where $n = n_1 + n_2, \mathbf{A} \in \mathbb{R}^{m \times n}, \mathbf{b} \in \mathbb{R}^m$, and $\mathbf{c} \in \mathbb{R}^n$. Let $\mathcal{J} \subseteq \{1, \ldots, n\}$ be a subset of the decision variables. The matheuristics investigated here work by fixing $x_j = \hat{x}_j$ for all $j \notin \mathcal{J}$ and solving the restricted MILP

$$\min \sum_{j \in \mathcal{J}} c_j x_j + \underbrace{\sum_{j \notin \mathcal{J}} c_j \hat{x}_j}_{\text{constant}} \tag{3.2a}$$

$$\text{s.t.} \sum_{j \in \mathcal{J}} a_{ij} x_j + \underbrace{\sum_{j \notin \mathcal{J}} a_{ij} \hat{x}_j}_{\text{constant}} \geq b_i \qquad \forall i = 1, \ldots, m \tag{3.2b}$$

$$\mathbf{x} \in \mathcal{X}(\mathcal{J}), \tag{3.2c}$$

where $\mathcal{X}(\mathcal{J}) = \{\mathbf{x} \in \mathbb{Z}^{n_1} \times \mathbb{R}^{n_2} : x_j = \hat{x}_j \text{ for all } j \notin \mathcal{J}\}$ is the neighborhood induced by $\mathcal{J}$. By judiciously selecting the set $\mathcal{J}$ of variables to be locally optimized, the induced neighborhoods (in this case, smaller, more tractable MILPs) can be searched more easily by a generic MILP solver. In other words, these matheuristics rely on "hard fixing" of variables to induce search neighborhoods. This is in contrast to, for example, local branching, a popular heuristic now available in all major MILP solvers, which searches restricted MILPs by adding "soft" constraints to (3.1) that allow at most some positive integer K variables to change from their current value [19]. This reliance on a single mathematical program is also in stark contrast to a branch-and-price algorithm in which two mathematical programs are maintained: one for the restricted master problem and one for the pricing subproblem. In the latter, if the problem statement changes, then both the master problem and pricing problems must be updated, debugged, and validated.

There are several noteworthy benefits of restricting ourselves to this class of matheuristics: (1) **Sustainment**. The issue of sustaining decision support software is rarely discussed in the academic domain, but is often a key factor in the acceptance, deployment, and success of a tool for an industrial application. It is typically easier for an optimization team to hand off a single mathematical programming model that relies on a general purpose solver, as opposed to low-level C code that may require an extended maintenance effort. This is particularly important as those who are tasked to support and maintain the software may change roles over time. (2) **Adapting to new business settings**. When the business environment and/or problem changes and new model features (e.g., constraints) are needed, it is preferable to update a single mathematical programming model as opposed to several models and the corresponding algorithms used to solve them. (3) **Handling extensions**. Deterministic optimization models can be extended to scenario-based stochastic programs with relative ease.

3.1.1 Literature Review

Brief review of matheuristics Archetti and Bertazzi [6] classify matheuristics for vehicle routing problems (a superset of IRPs) into three classes, which we state verbatim:

1. *Decomposition approaches*. In general, in a decomposition approach the problem is divided into smaller and simpler subproblems and a specific solution method is applied to each subproblem. In matheuristics, some or all these subproblems are solved through mathematical programming models to optimality or suboptimality.
2. *Improvement heuristics*. Matheuristics belonging to this class use mathematical programming models to improve a solution found by a different heuristic approach. They are very common as they can be applied whatever heuristic is used to obtain a solution that the mathematical programming model aims at improving.
3. *Branch-and-price/column generation-based approaches*. Branch-and-price algorithms have been widely and successfully used for the solution of routing problems. Such algorithms make use of a set partitioning formulation, where a binary or integer variable is associated with each possible route (column). Due to the exponential number of variables, the solution of the linear relaxation of the formulation is performed through column generation. In the branch-and-price/column generation-based matheuristics the exact method is modified to speed up the convergence, thus losing the guarantee of optimality. For example, the column generation phase is stopped prematurely.

In this chapter, we focus exclusively on matheuristics in the first two categories.

Brief review of MIRPs Papageorgiou et al. [36] provide a detailed survey of MIRPs with inventory tracking at all ports, i.e., MIRPs in which inventory levels at all loading and discharging ports are required to remain within prespecified bounds during every time period of the planning horizon. To avoid significant overlap with that review, we discuss some applications and recent work in this section. We discuss several algorithmic papers in Sect. 3.3 when we discuss the matheuristics compared in this study so that highly relevant methods are discussed in context.

Several notable case studies involving real-world maritime inventory routing applications have appeared in the literature. In collaboration with a Norwegian calcium carbonate slurry supplier, Dauzère-Pérès et al. [17] describe a case study in VMI in which the supplier is in charge of routing a fleet of heterogeneous vessels throughout Northern Europe and of ensuring sufficient inventory levels of up to sixteen products at ten tank farms. Christiansen et al. [13] present a MIRP faced by a large cement producer involving bulk ships possessing multiple compartments that transport multiple non-mixable cement products. In the petrochemicals sector, Furman et al. [21] demonstrate the benefits of an arc-flow MILP model within a decision support tool used to assist decision-makers make tactical routing and inventory management decisions for vacuum gas oil at ExxonMobil. This tramp shipping application is an important example of a single product MIRP involving vessels that are chartered for a single voyage between Northwest Europe and the Gulf Coast.

To solve models akin to those described in [21], Song and Furman [42] build upon the techniques introduced in Savelsbergh and Song [40] by applying a large neighborhood search algorithm to an arc-flow model. Specifically, after they generate an initial solution, they apply a local search heuristic resembling a 2-opt procedure in which an exact optimization method locally optimizes the decision variables associated with all but two vessels, while all other vessel decisions are fixed. They apply this procedure at most $\binom{|\mathcal{V}|}{2}$ iterations, where $|\mathcal{V}|$ denotes the number of vessels, and randomly select vessel pairs in each iteration. Engineer et al. [18] tackle a less operationally complex problem than the one considered in [21, 42]. They formulate the problem using a path-flow model and solve it via a branch-cut-and-price algorithm that makes use of three types of valid inequalities. In Hewitt et al. [30], branch-and-price guided search (BPGS) is used as a primal heuristic to find good solutions quickly for the instances considered in [18]. Exploiting information of a master problem's corresponding extended formulation, BPGS systematically searches restricted neighborhoods of a MILP [29]. In addition to introducing more comprehensive local search neighborhoods than previously studied, Hewitt et al. [30] demonstrate that their parallel BPGS implementation effectively identifies high-quality solutions in under 30 min for the MIRP instances considered.

With few exceptions, MIRPs appearing in the literature review can be cast as mixed-integer linear programs (MILPs). In general, as the number of components (e.g., vessels, ports, and time periods) increase so too does the computation time required to solve the corresponding problem instances. Worse, there is often a point at which attempting to solve a realistic instance using a commercially available solver yields unacceptable performance, e.g., a low quality solution in a given time limit. To combat these limitations, numerous heuristics have been developed that exploit mathematical programming solvers.

In the liquefied natural gas (LNG) domain, several heuristics for LNG-IRPs have been studied. Rakke et al. [38] propose a rolling horizon heuristic in which a sequence of overlapping MILP subproblems are solved. Each subproblem involves at most 3 months of data and consists of a 1-month "central period" and a "forecasting period" of at most 2 months. Once a best solution is found (either by optimality or within a time limit), all decision variables in the central period are fixed at their respective values and the process "rolls forward" to the next subproblem. Stålhane et al. [43] propose a construction and improvement heuristic that builds voyage schedules based on the availability of vessels and product while simultaneously satisfying inventory constraints. Goel et al. [24] also use construction and improvement heuristics within a large neighborhood search. To address huge instances, Asokan et al. [7] demonstrate how these neighborhood searches can be parallelized to significantly reduce computation time. Goel et al. [25] introduce a constraint programming approach for the same LNG-IRP that can achieve an order of magnitude speedup relative to CPLEX 11.1. Mutlu et al. [33] employ a vessel routing heuristic for an LNG-IRP that iteratively builds ship routes that ensure delivery times and volume requirements of each contract are met.

Solution techniques for MIRPs faced by a vertically integrated company are presented in [36], where MIRPs with inventory tracking at every port are surveyed. Fodstad et al. [20] directly apply a commercial MILP solver, while Uggen et al. [47] employ a fix-and-relax primal heuristic. Adapting Song and Furman's [42] local search heuristic, Goel et al. [24] describe a simple construction heuristic to identify good solutions to large MIRP instances with up to 365 time periods. Their model seeks to minimize penalties and does not consider travel costs.

More recently, several papers have appeared on short-sea shipping applications. Agra et al. [2] consider and combine three heuristic procedures for a fuel oil distribution problem. A rolling horizon heuristic is used to decompose the original problem into smaller and more tractable problems, feasibility pump is used to find initial solutions for MILP problems, and local branching is used to improve feasible solutions. Hemmati et al. [28] evaluated the potential of having a VMI service in tramp shipping to provide flexibility in delivery time and cargo quantities. Hemmati et al. [27] propose a two-phase hybrid matheuristic to solve a multi-product short sea MIRP. They first convert the MIRP into a ship routing and scheduling problem and then apply an adaptive large neighborhood search to solve the resulting optimization problem. Jiang and Grossmann [31] evaluate a number of MILP formulations for an operational MIRP; no heuristics are applied.

3.1.2 Contributions of this Chapter

The contributions of this chapter are:

1. Several variants of popular matheuristics used in the MIRP literature are presented and adapted to a particular class of long-horizon MIRPs.
2. An extensive computational study comparing a suite of simple matheuristics that rely on the same underlying mathematical programming formulation is performed.
3. New best known incumbents are found for 26 out of 70 yet-to-be-proved-optimal instances and new best known bounds on 56 instances.

The remainder of the chapter is organized as follows: In Sect. 3.2, we present a discrete-time arc-flow formulation of the problem. In Sect. 3.3, we describe the matheuristics used in our computational study, while citing other works in which they have been used. Section 3.4 includes a detailed computational comparison of these matheuristics against one another as well as against commercial MILP solvers CPLEX 12.6.2 and Gurobi 6.5. Concluding remarks and suggestions for future research directions are presented in Sect. 3.5.

3.2 Problem Description and Formulations

We consider a single product MIRP that involves a planning horizon of T time periods, where $\mathcal{T} = \{1, \dots, T\}$ is the set of all time periods in the horizon, as well as three main components: ports, vessels, and vessel classes. Each port is classified as a loading port (where product is produced) or as a discharging port (where product is consumed). Let $\mathcal{J}^P$ denote the set of loading (production) ports, $\mathcal{J}^C$ the set of consumption (or discharging) ports, and $\mathcal{J} = \mathcal{J}^P \cup \mathcal{J}^C$ the set of all ports. Let the parameter Δ_j be 1 if $j \in \mathcal{J}^P$ and -1 if $j \in \mathcal{J}^C$. Product can be stored in inventory at both types of ports. Each port has: an inventory capacity of $S^{\max}_{j,t}$; a fixed number of berths B_j restricting the number of vessels that can simultaneously load or discharge in a given period; and a deterministic, but possibly non-constant, per-period rate $D_{j,t}$ of production or consumption, i.e., $D_{j,t}$ denotes the amount of product produced or consumed at port j in time period t. The amount of inventory at the end of time period t must be between 0 and $S_{j,t}^{\max}$.

Since it may not be possible to satisfy all demand or avoid hitting tank top, we include a simplified spot market so that a consumption port may buy product and a production port can sell excess inventory whenever necessary. The penalty parameter $P_{j,t}$ denotes the unit cost associated with the spot market at port j in time period t. We assume that $P_{j,t} > P_{j,t+1}$ for all $t \in \mathcal{T}$ so that the spot market is only used as late as possible, i.e., to ensure that a solution will not involve lost production (stockout) until the inventory level reaches capacity (falls to zero). When the penalty parameters are large (like a traditional "Big M" value), inventory bounds can be considered "hard" constraints. When they are small, however, inventory bounds can be treated as "soft" constraints. This "soft" interpretation may be beneficial in strategic planning problems for several reasons. First, a user may attempt to solve an instance with a demand forecast that cannot be met by the existing fleet in order to understand the limitations of the current infrastructure (see also [24]). Second, the inventory bounds given as input may be overly conservative in order to make the solution more robust when, in fact, slight bound violations may be acceptable. Third, incurring a small penalty for a particular solution (as opposed to declaring it strictly infeasible) can mitigate minor unwanted effects of using a discrete-time model [36].

Vessels travel from port to port, loading and discharging product. We assume vessels fully load and fully discharge at a port and that direct deliveries are made. Each vessel belongs to a vessel class $vc \in \mathcal{VC}$. Vessel class vc has capacity Q^{vc}. Vessels are owned by the supplier or time-chartered for the entire planning horizon. We assume that port capacity always exceeds vessel capacity, i.e., $S_{j,t}^{\max} \geq \max\{Q^{vc} : vc \in \mathcal{VC}\}$ and that vessels can fully load or discharge in a single period. These assumptions allow vessels to load or discharge in the same period in which they leave a port so that loading and discharging decisions do not need to be explicitly modeled.

As is done in several other works, the problem is modelled on a time-expanded network. The network has a set $\mathcal{N}_{0,T+1}$ of nodes and a set $\mathcal{A}$ of directed arcs. The node set is shared by all vessel classes, while each vessel class has its own arc set $\mathcal{A}^{vc}$. The set $\mathcal{N}_{0,T+1}$ of nodes consists of a set $\mathcal{N} = \{(j,t) : j \in \mathcal{J}, t \in \mathcal{T}\}$ of "regular" nodes, or port-time pairs, as well as a source node n_0 and a sink node n_{T+1}.

Associated with each vessel class vc is a set $\mathcal{A}^{vc}$ of arcs, which can be partitioned into source, sink, waiting, and travel arcs. A source arc $a = (n_0, (j,t))$ from the source node to a regular node represents the arrival of a vessel to its initial destination. A sink arc $a = ((j,t), n_{T+1})$ from a regular node to the sink node indicates that a vessel has departed the system and is no longer in use. A waiting arc $a = ((j,t),(j,t+1))$ from a port j in time period t to the same port in time period $t+1$ represents a vessel remaining at that port in two consecutive time periods. Finally, a travel arc $a = ((j_1,t_1),(j_2,t_2))$ with $j_1 \neq j_2$ represents travel between two distinct ports, where the travel time $(t_2 - t_1)$ between ports is given. If a travel or sink arc is taken, we assume that a vessel fully loads or discharges immediately before traveling. The cost of traveling on arc $a \in \mathcal{A}^{vc}$ is C_a^{vc}.

The set of all travel and sink arcs for each vessel class are denoted by $\mathcal{A}^{vc,\text{inter}}$ (where "inter" stands for "inter-regional"). The set of incoming and outgoing arcs associated with vessel $vc \in \mathcal{VC}$ at node $n \in \mathcal{N}_{0,T+1}$ are denoted by $\mathcal{RS}_n^{vc}$ (for reverse star) and $\mathcal{FS}_n^{vc}$ (for forward star), respectively. Similarly, $\mathcal{FS}_n^{vc,\text{inter}}$ denotes the set of all outgoing travel and sink arcs at node n for vessel class vc. For our strategic planning problem, modeling the flow of vessel classes avoids the additional level of detail associated with modeling each individual vessel. Moreover, we found that modeling vessel classes could remove symmetry and improve solution times by more than an order of magnitude on large instances.

Assumptions: For ease of reference, we collect the assumptions made throughout this chapter: (1) There is exactly one port within each region; (2) Port capacity always exceeds the capacity of the vessels, e.g., $S_{j,t}^{\max} \geq \max\{Q^{vc} : vc \in \mathcal{VC}\}$; (3) Travel times are deterministic; (4) Vessels can fully load or discharge in a single period (in other words, the time to load/discharge is deterministic and built into the travel time); (5) Production and consumption rates are known; (6) There is a single loading port ($|\mathcal{J}^P| = 1$) as is typically the case for LNG-IRPs [24, 26, 37, 43]; (7) In a single time period, at most one vessel per vessel class may begin an outgoing voyage to a given discharging port or a return voyage to the loading port. The reason for this assumption is that, in practice, a strategic planner is interested in avoiding scheduling conflicts at a port when a plan is actually enacted. Said differently, a schedule in which two large vessels of the same class must load or discharge simultaneously at the exact same port is perceived to be a non-robust solution.

3.2.1 A Discrete-Time Arc-Flow Mixed-Integer Linear Programming Model

We next define the decision variables. Let x_a^{vc} be the number of vessels in vessel class vc that travel on arc $a \in \mathcal{A}^{vc}$. Let $s_{j,t}$ be the ending inventory at port j in time period t. Initial inventory $s_{j,0}$ is given as data. Finally, let $\alpha_{j,t}$ be the amount of inventory bought from or sold to the spot market near port j in time period t.

We consider the following discrete-time arc-flow MILP model, which was introduced in Papageorgiou et al. [34]:

$$\min \sum_{vc \in \mathcal{VC}} \sum_{a \in \mathcal{A}^{vc}} C_a^{vc} x_a^{vc} + \sum_{j \in \mathcal{J}} \sum_{t \in \mathcal{T}} P_{j,t} \alpha_{j,t} \tag{3.3a}$$

$$\text{s.t.} \quad \sum_{a \in \mathcal{FS}_n^{vc}} x_a^{vc} - \sum_{a \in \mathcal{RS}_n^{vc}} x_a^{vc} = \begin{cases} +1 & \text{if } n = n_0 \\ -1 & \text{if } n = n_{T+1} \\ 0 & \text{if } n \in \mathcal{N} \end{cases}, \qquad \forall\, n \in \mathcal{N}_{0,T+1}, vc \in \mathcal{VC} \tag{3.3b}$$

$$s_{j,t} = s_{j,t-1} + \Delta_j \left(D_{j,t} - \sum_{vc \in \mathcal{VC}} \sum_{a \in \mathcal{FS}_n^{vc,\text{inter}}} Q^{vc} x_a^{vc} - \alpha_{j,t} \right), \qquad \forall\, n = (j,t) \in \mathcal{N} \tag{3.3c}$$

$$\sum_{vc \in \mathcal{VC}} \sum_{a \in \mathcal{FS}_n^{vc,\text{inter}}} x_a^{vc} \leq B_j, \qquad \forall\, n = (j,t) \in \mathcal{N} \tag{3.3d}$$

$$\alpha_{j,t} \geq 0, \qquad \forall\, n = (j,t) \in \mathcal{N} \tag{3.3e}$$

$$s_{j,t} \in [0, S^{\max}_{j,t}] , \qquad \forall\, n = (j,t) \in \mathcal{N} \tag{3.3f}$$

$$x^{vc}_a \in \{0,1\} , \qquad \forall\, vc \in \mathcal{VC} ,\, a \in \mathcal{A}^{vc,\text{inter}} \tag{3.3g}$$

$$x^{vc}_a \in \mathbb{Z}_+ , \qquad \forall\, vc \in \mathcal{VC} ,\, a \in \mathcal{A}^{vc} \setminus \mathcal{A}^{vc,\text{inter}} . \tag{3.3h}$$

The objective is to minimize the sum of all transportation costs and penalties for lost production and stockout. Constraints (3.3b) require flow balance of vessels within each vessel class. Constraints (3.3c) are inventory balance constraints at loading and discharging ports, respectively. Berth limit constraints (3.3d) restrict the number of vessels that can attempt to load/discharge at a port at a given time. This formulation requires that a vessel must travel at capacity from a loading region to a discharging region and empty on the return voyage to a loading region. This model does not require decision variables for tracking inventory on vessels (vessel classes), nor does it include decision variables for the quantity loaded/discharged in a given period.

This model is similar to the one studied in Goel et al. [24]. The major differences are that they do not include travel costs in the objective function; they model each vessel individually (in other words, there is only one vessel per vessel class); they model consumption rates as decision variables with upper and lower bounds; and they include an additional set of continuous decision variables to account for cumulative unmet demand at each consumption port.

3.2.2 Additional Valid Inequalities: Mixed Integer Rounding Cuts

It is unclear if additional constraints are beneficial for finding better primal solutions. On the one hand, more constraints typically imply that a solver must do more work at each search tree node, e.g., during each dual simplex update since the constraint matrix $\mathbf{A}$ is larger. On the other hand, a tighter relaxation often reduces the number of nodes that must be explored. Below we describe a set of mixed integer rounding cuts that can be added to the formulation. These cuts were originally proposed in Engineer et al. [18] and later in Agra et al. [1] and Papageorgiou et al. [36]. For completeness, a formal derivation is provided in the appendix.

Let $R_{j,t}$ be the cumulative amount of product required for pick up or delivery at port j by time period t, i.e.,

$$R_{j,t} = \begin{cases} \sum_{t' \le t} D_{j,t'} - s_{j,0} & \text{if} \quad j \in \mathcal{J}^{\mathcal{C}}, \\ \sum_{t' \le t} D_{j,t'} + s_{j,0} - S^{\max}_j & \text{if} \quad j \in \mathcal{J}^{\mathcal{P}}. \end{cases} \tag{3.4}$$

Then, for any scalar $Q > 0$, the following mixed integer rounding cuts are valid:

$$\sum_{vc \in \mathcal{VC}} \sum_{t' \le t} \sum_{a \in \mathcal{FS}^{vc,\text{inter}}_{j,t'}} \left(\left\lceil \frac{Q^{vc}}{Q} \right\rceil - \frac{(F_{vc} - F^0_{j,t})^+}{1 - F^0_{j,t}} \right) x^{vc}_a + \sum_{t' \le t} \frac{\alpha_{j,t'}}{Q(1 - F^0_{j,t})} \ge \left\lceil \frac{R_{j,t}}{Q} \right\rceil \quad \forall (j,t) \in \mathcal{N}, \tag{3.5}$$

where $F^0_{j,t} = \frac{R_{j,t}}{Q} - \left\lceil \frac{R_{j,t}}{Q} \right\rceil$, $F_{vc} = \frac{Q^{vc}}{Q} - \left\lceil \frac{Q^{vc}}{Q} \right\rceil$. These constraints are valid as they result from a direct application of the mixed-integer rounding cut (3.11) (see the appendix) applied to the constraint

$$\sum_{vc \in \mathcal{VC}} \sum_{t' \leq t} \sum_{a \in \mathcal{FS}^{vc,\text{inter}}_{j,t'}} \frac{Q^{vc}}{Q} x^{vc}_a + \frac{1}{Q} \sum_{t' \leq t} \alpha_{j,t'} \geq \frac{R_{j,t}}{Q} \quad \forall (j,t) \in \mathcal{N},$$

for any $Q > 0$. In our experiments, we set $Q = Q^{\max} = \max\{Q^{vc} : vc \in \mathcal{VC}\}$. Of course, one could choose several values of Q to derive an assortment of cuts. It is not yet known if including a vast array of problem-specific cuts will outperform the myriad of general-purpose cuts already available in a MILP solver's cut library.

Although constraints (3.5) only consider cumulative supply and demand in a time interval $[0, t]$, it is possible to include such constraints for any time interval $[t_1, t_2]$ for any $t_1, t_2 \in \mathcal{T}$ such that $t_1 < t_2$. To do so, one must define

$$R_{j,t_1,t_2} = \begin{cases} D_{j,[t_1,t_2]} - S^{\max}_{j,t_1-1} + S^{\min}_{j,t_2} & \text{if} \quad j \in \mathcal{J}^C, \\ D_{j,[t_1,t_2]} + S^{\min}_{j,t_1-1} - S^{\max}_{j,t_2} & \text{if} \quad j \in \mathcal{J}^P, \end{cases}$$

where $D_{j,[t_1,t_2]} = \sum_{t'=t_1}^{t_2} D_{j,t'}$, and $S^{\min}_{j,t}$ and $S^{\max}_{j,t}$ are known or improved bounds on $s_{j,t}$. This is particularly relevant when implemented within a rolling horizon approach since s_{j,t_1-1} will be known when t_1 is the starting time period of the interval being optimized. In this case, we have $S^{\min}_{j,t_1-1} = S^{\max}_{j,t_1-1} = s_{j,t_1-1}$.

3.3 Matheuristics Explored

3.3.1 Construction Heuristics

Reducing the number of time periods Aggregation in space and/or time are commonplace when attempting to reduce the solution space of a large optimization problem. In this approach, we attempt to reduce the number of time periods in which vessels can travel between distinct ports. In our implementation, rather than manipulate the set $\mathcal{T}$ of time periods, we limited the set of travel arcs (i.e., arcs that involve two distinct ports) so that vessels may only travel in even-numbered periods. The berth limit constraints were not modified. We attempted several other variants, but did not see any benefits for this class of problems.

Rolling horizon heuristics Arguably the most common form of time decomposition in engineering applications is some form of rolling horizon heuristic (RHH). In the IRP literature, Al-Ameri et al. [3] present a rolling horizon framework for vendor-managed inventory systems. In the MIRP literature, this approach was applied in Al-Khayyal and Hwang [4], Rakke et al. [38], Agra et al. [2], and others. The basic idea of an RHH tailored to time-based IRPs is to solve a sequence of overlapping

MILP subproblems. Each subproblem consists of a "central period" and a "forecasting period." Once an optimal solution to this subproblem is found (or sub-optimal solution within a time limit), all decision variables in the central period are fixed at their respective values and the process "rolls forward" to the next subproblem.

Our implementation is shown in Algorithm 1. Here τ^{central} and τ^{forecast} are the number of periods in the central and forecast periods, respectively. The set $\mathcal{T}^{\text{start}} = \{t \in \mathcal{T} : t = 1 + \nu\tau^{\text{central}} \text{ for } \nu \in \mathbb{N}\}$ denotes the set of all starting periods to consider in the procedure. The algorithm is initialized by forcing all product to be bought and sold from the spot market $\alpha_{j,t} = D_{j,t}$ for all (j, t), which implies that inventory remains constant $s_{j,t} = s_{j,t-1}$ at all ports over the entire planning horizon. Meanwhile, all vessels movements are initially forbidden, i.e., $x_a^{vc} = 0$ for all $vc \in \mathcal{VC}$, $a \in \mathcal{A}^{vc}$. Note that this means that vessel balance constraints (3.3b) are not feasible at the source node and the initial solution is infeasible. This infeasibility is immediately resolved within the first solve. The algorithm is straightforward except for one subtle feature not common in all rolling horizon heuristics: In Step 1, in all but the final solve, it is wise for this application to forbid vessels to exit the system in the central period. Without this fixing, it may be optimal with a myopic forecast to have some vessels exit the system (take a sink arc) in the central period, which will in turn make these vessels unavailable in all future periods even after the algorithm rolls forward.

Algorithm 1 Rolling Horizon Heuristic

1: Initialization: Set $\alpha_{j,t} = D_{j,t}\ \forall(j,t)$; $s_{j,t} = s_{j,t-1}\ \forall(j,t)$; $x_a^{vc} = 0\ \forall vc \in \mathcal{VC}, a \in \mathcal{A}^{vc}$
2: **for** $t^{\text{start}} \in \mathcal{T}^{\text{start}}$ **do**
3: $t^{\text{end}} = \min\{t^{\text{start}} + \tau^{\text{central}} + \tau^{\text{forecast}}, T\}$
4: Fix all decision variables with index $t < t^{\text{start}}$ or $t > t^{\text{end}}$ to their current value
5: Deactivate all constraints with index $t < t^{\text{start}}$ or $t > t^{\text{end}}$
6: If $\texttt{Ord}(t^{\text{start}}) < \texttt{Card}(\mathcal{T}^{\text{start}})$, fix all sink arc variables $x^{vc}_{((j,t),n_{T+1})} = 0$ for all $t \leq t^{\text{start}} + \tau^{\text{central}}$
7: Solve Model (3.3)
8: **end for**

Several variants are worth mentioning. Uggen et al. [47] present a fix-and-relax heuristic in which the forecast period includes all remaining time periods and all integer decision variables in the forecast period are relaxed to be continuous. This did not work well for this class of problems, so we did not adopt it. Goel et al. [24] and Shao et al. [41] apply a rolling horizon framework within a construction heuristic, but without solving a mathematical program. From a high level, the approximate dynamic programming method proposed in Papageorgiou et al. [34], the first of its kind for maritime inventory routing problems, can be viewed as a type of rolling horizon heuristic that uses value functions to approximate the value of sending vessels to certain ports in the forecast period. It is more complex than the matheuristics evaluated in this work. Similarly, Toriello et al. [46] apply approximate dynamic programming to address a deterministic IRP. It is one of the first papers to apply value function approximations to an IRP, extending a body of work on dynamic fleet management problems [22, 23, 44, 45].

Restricting port-vessel class compatibilities For instances with many ports and vessel classes, Model (3.3) requires many binary variables, namely, one for each tuple $(vc, j_1, t, j_2, t + \tau_{j_1,j_2})$ where τ_{j_1,j_2} is the travel time between ports j_1 and j_2. If one were to restrict certain vessel class from traveling to certain discharging ports, then fewer binary decisions would be required. The approach that follows attempts to do just that in a systematic manner. We formulate a MILP whose purpose is to identify a set of port-vessel class incompatibilities that lead to a restricted version of Model (3.3) that is easier to solve, but still generates high-quality solutions. Interestingly, this approach could be viewed as a type of a "Cluster first—route second" approach discussed in [6].

The basic idea behind the MILP below is to solve a much simplified variant of Model (3.3) with the requirement that at least $K \in \mathbb{N}$ port-vessel class pairs are incompatible so that a subset of less promising travel arcs are eliminated. Unlike Model (3.3), this MILP models each individual vessel, but does not involve any sequencing. It exploits the fact that there is only one supply port by assuming all vessels originate at the supply port. For those vessels that originate at a demand port, we assume that they can discharge in the first period in which they are available so that they then travel immediately to the supply port. This travel time is then deducted from the vessel's remaining time available to service other ports. Waiting time (i.e., successive time periods spent at a port) is not modeled. The model is described below.

Indices and Sets

$t \in \mathcal{T}'$ subset of time periods considered; we assume $T \in \mathcal{T}'$
$v \in \mathcal{V}$ set of individual vessels

Parameters

$T^{\text{Rem}}_{t,v}$ remaining time available up to time t for vessel v from its first possible arrival to the supply port
$T^{\text{OW}}_{j,v}$ one-way travel time from the supply port to demand port j for vessel v
$T^{\text{RT}}_{j,v}$ roundtrip travel time from the supply port to demand port j (and back) for vessel v
$C^{\text{OW}}_{j,v}$ travel cost of a one-way voyage from the supply port to demand port j for vessel v
$C^{\text{RT}}_{j,v}$ travel cost of a roundtrip voyage from the supply port to demand port j (and back) for vessel v
$D^{\text{LB}}_{j,t}$ cumulative excess demand at demand port j up to time period t
$D^{\text{UB}}_{j,t}$ maximum demand at demand port j up to time period t
P_j penalty for unmet demand: $P_j = P_{j,1}$, for all j

The parameters $T^{\text{Rem}}_{t,v}$, $D^{\text{LB}}_{j,t}$, and $D^{\text{UB}}_{j,t}$ are set as follows: For a vessel originating at the supply port, $T^{\text{Rem}}_{t,v} = [t - T^{\text{start}}_v]^+$. For a vessel originating at a discharging port, we set $T^{\text{Rem}}_{t,v} = [t - T^{\text{start}}_v - T^{\text{OW}}_{j(v),v}]^+$, where $j(v)$ denotes the discharging port where vessel v originates. This is an optimistic value since the vessel may not be able to fully discharge in the same period in which it enters the system/network. We set $D^{\text{LB}}_{j,t} = R_{j,t} - \bar{S}_{j,t}$, where $R_{j,t}$, defined in (3.4), is the cumulative amount of product required to be delivered by time period t defined in (3.4) and $\bar{S}_{j,t} = \sum_{v \in \mathcal{V}: j(v)=j, t \leq T^{\text{start}}_v} Q_v$ is the expected amount discharged by vessels originating at discharging port j by time t before making their first voyage to the supply port. Meanwhile, $D^{\text{UB}}_{j,t} = D^{\text{LB}}_{j,t} + S^{\max}_j$. All other parameters listed above are given as data.

Decision Variables

$\sigma_{j,t}$ (continuous) cumulative slack at demand portdemand port j up to time t

$x^{\text{OW}}_{j,t,v}$ (binary) cumulative number of one-way voyages to demand port j up to time period t made by vessel v

$x^{\text{RT}}_{j,t,v}$ (integer) cumulative number of roundtrip voyages to demand port j up to time period t made by vessel v

$z_{j,vc}$ (binary) takes value 1 if demand portdemand port j and vessel class vc are deemed incompatible; 0 otherwise

Port-Vessel Class Incompatibility MIP Model

$$\min \sum_{j \in \mathcal{J}^C} \sum_{v \in \mathcal{V}} \left(C^{\text{RT}}_{j,v} x^{\text{RT}}_{j,T,v} + C^{\text{OW}}_{j,v} x^{\text{OW}}_{j,T,v} \right) + \sum_{j \in \mathcal{J}^C} P_j \sigma_{j,T} \tag{3.6a}$$

$$\text{s.t.} \sum_{j \in \mathcal{J}^C} T^{\text{RT}}_{j,v} x^{\text{RT}}_{j,t,v} + T^{\text{OW}}_{j,v} x^{\text{OW}}_{j,t,v} \leq T^{\text{Rem}}_{t,v}, \quad \forall t \in \mathcal{T}', v \in \mathcal{V} \tag{3.6b}$$

$$\sum_{v \in \mathcal{V}} Q^v (x^{\text{RT}}_{j,t,v} + x^{\text{OW}}_{j,t,v}) + \sigma_{j,t} \geq D^{\text{LB}}_{j,t}, \quad \forall j \in \mathcal{J}^C, t \in \mathcal{T}' \tag{3.6c}$$

$$\sum_{v \in \mathcal{V}} Q^v (x^{\text{RT}}_{j,t,v} + x^{\text{OW}}_{j,t,v}) + \sigma_{j,t} \leq D^{\text{UB}}_{j,t}, \quad \forall j \in \mathcal{J}^C, t \in \mathcal{T}' \tag{3.6d}$$

$$x^{\text{RT}}_{j,t,v} + x^{\text{OW}}_{j,t,v} \geq x^{\text{RT}}_{j,t-1,v} + x^{\text{OW}}_{j,t-1,v}, \quad \forall j \in \mathcal{J}^C, t \in \mathcal{T}' \tag{3.6e}$$

$$\sigma_{j,t} \geq \sigma_{j,t-1}, \quad \forall j \in \mathcal{J}^C, t \in \mathcal{T}' \tag{3.6f}$$

$$T^{\text{RT}}_{j,v} x^{\text{RT}}_{j,T,v} + T^{\text{OW}}_{j,v} x^{\text{OW}}_{j,T,v} \leq T^{\text{Rem}}_{T,v}(1 - z_{j,vc}), \quad \forall j \in \mathcal{J}^C, vc \in \mathcal{VC}, v \in \mathcal{V} : VC(v) = vc \tag{3.6g}$$

$$\sum_{j,vc} z_{j,vc} \geq K, \tag{3.6h}$$

$$\sigma_{j,t} \geq 0, \quad \forall j \in \mathcal{J}^C, t \in \mathcal{T}' \tag{3.6i}$$

$$x^{\text{RT}}_{j,t,v} \in \mathbb{Z}_+, \quad \forall j \in \mathcal{J}^C, t \in \mathcal{T}', v \in \mathcal{V} \tag{3.6j}$$

$$x^{\text{OW}}_{j,t,v} \in \{0, 1\}, \quad \forall j \in \mathcal{J}^C, t \in \mathcal{T}', v \in \mathcal{V} \tag{3.6k}$$

$$z_{j,vc} \in \{0, 1\}, \quad \forall j \in \mathcal{J}^C, vc \in \mathcal{VC}. \tag{3.6l}$$

The objective is to minimize the total cost of one-way and roundtrip voyages and cumulative unmet demand penalties up to time T. Constraints (3.6b) limit the number of one-way and roundtrip voyages each vessel can make. Constraint (3.6h) requires at least K discharging port-vessel class pairs to be incompatible. Constraints (3.6c) and (3.6d) bound the amount of product that can be discharged at each discharging port up to time $t \in \mathcal{T}'$. Note that the more periods considered (i.e., the larger $\mathcal{T}'$ is), the more difficult it becomes to solve this MILP to provable optimality. For example, if $\mathcal{T}' = \mathcal{T}$, then this MILP would be almost as difficult to solve as Model (3.3). Constraints (3.6e) ensure that the cumulative number of voyages is correct. Note that it allows for a single one-way voyage to be made up to time period $t - 1$ (e.g., $x^{\text{OW}}_{j,t-1,v} = 1$), but then the return trip to be completed after time period $t - 1$ such that no one-way trip is made up to time period t (e.g., $x^{\text{OW}}_{j,t,v} = 0$).

Since the interplay of one-way and roundtrip voyages is important, the following example is meant to clarify how the two are used in Model (3.6). Consider an instance with a single supply port, a single demand port, and a single vessel originating at the supply port and available in the first time period. Assume a 181-day planning horizon with $\mathcal{T} = \{0, 1, \ldots, 180\}$. Assume that a one-way (roundtrip) voyage requires 10 (20) days. Assume that, in an optimal solution, the demand port is visited in time periods t^* for $t^* \in \{10, 30, 50, 70, 90, 110, 130, 150, 170\}$. Now, let $\mathcal{T}' = \{90, 180\}$ be the subset of time periods considered in Model (3.6). Omitting the subscripts j and v in $x^{\text{RT}}_{j,t,v}$ and $x^{\text{OW}}_{j,t,v}$ (since there is only one demand port and one vessel), this optimal solution corresponds to a solution $x^{\text{RT}}_{90} = 4$, $x^{\text{OW}}_{90} = 1$, $x^{\text{RT}}_{180} = 8$, and $x^{\text{OW}}_{180} = 1$. That is, up to time period 90, there are four roundtrip voyages in which the vessel delivers in time periods $t^* \in \{10, 30, 50, 70\}$ and a single one-way voyage in which the vessel delivers in time period 90. Up to time period 180, there are eight roundtrip voyages and a single one-way voyage. Note that if one-way voyages were only permitted in the last time period (i.e., $t = 180$ in this example), only four roundtrips would have been feasible up to time period 90 possibly causing slack σ_{90} to be positive. Since $\sigma_{j,t}$ denotes *cumulative* slack, an incorrect amount of slack would have been incurred.

Parameter settings: In our computational experiments, we set $K = 0.2|\mathcal{J}^C||\mathcal{VC}|$ and $\mathcal{T}' = \{t \in \mathcal{T} : t \bmod T/T' = 0\}$ with $T' = 12$ so that $T' = |\mathcal{T}'| = 12$.

3.3.2 *Improvement Heuristics*

Below we discuss several matheuristics that are commonly used as improvement heuristics, i.e., as tools to improve existing feasible, or perhaps infeasible solutions.

***K*-opt local search** Borrowing from the K-opt local search procedures commonly used in combinatorial optimization problems, this heuristic fixes the routes for all but K vessel classes, and solves MILP (3.3) for a better solution over the neighborhood defined by the K unfixed vessel classes. This matheuristic is remarkably simple to implement and was used in [24, 30, 34, 35, 42], although often with routes for vessels, rather than vessel classes, being fixed and optimized. Pseudocode is provided in Algorithm 2.

Algorithm 2 K-opt Heuristic

1: **while** `elapsedTime` $\leq$ `timeLimit and improvementOccurred` **do**
2: **for** $\mathcal{VC}' \in \mathcal{S}$ **do**
3: Fix all decision variables associated with vessels in vessel class $vc \notin \mathcal{VC}'$ to their current value
4: Solve Model (3.3)
5: **end for**
6: **end while**

The most interesting step in the algorithm is deciding which vessel classes should be fixed and in what order. In this chapter, we simply use a predefined ordering so that the set $\mathcal{S}$ of vessel class tuples is given as input. Goel et al. [24] suggested a tailored scheme for dynamically choosing pairs of vessels to include in a 2-opt approach for an LNG-IRP. It is important to note that the order in which vessel classes are selected can have a significant impact on the performance of the algorithm.

1-opt with D-day time window flexibility search This algorithm is a variant of the time-window improvement heuristic introduced in Goel et al. [24] and used subsequently in Shao et al. [41]. As proposed in [24], this heuristic searches a neighborhood consisting of all solutions such that any voyage departure is delayed or advanced by at most D days, relative to the departure date for that voyage in the incumbent solution. We extend this heuristic by increasing the size of the neighborhood to include a `1opt` search. Specifically, in addition to allowing voyages $2D$ days of additional flexibility (D days before and D days after), we also unfix all variables associated with a single vessel class. Thus, this heuristic searches over larger neighborhoods than `1opt`.

Solution pool-based polishing This algorithm seeks to improve the best known solution given a pool of solutions by attempting to exploit the best properties of these solutions. It is similar in spirit to that of [39]. The method solves an MILP in which travel and demurrage arcs are active (unfixed) only if they were selected by one of the solutions in the solution pool. All other travel and demurrage arcs are removed from (fixed to zero in) the problem formulation. The MILP is initialized with the best solution from the pool, and can be terminated by achieving proven optimality or by reaching the time limit. Unlike other heuristic methods, this heuristic can take multiple solutions as input.

The solution pool can be generated by MILP solvers, or by saving the solutions found through different approaches. This heuristic can be used as a last step to improve the solutions found with other methods.

3.3.3 Hybrid Approaches

Combining the construction and improvement heuristics presented above has the potential to further improve solution quality. Several hybridizations are discussed below. Before proceeding, we pause to mention that it is not clear a priori if solving each portion of a rolling horizon heuristic to provable optimality is, indeed, better than solving a portion to near optimality before rolling forward. For example, an optimal solution to a myopic instance may place vessels in a suboptimal future position from which it is impossible to recover.

Rolling horizon heuristic with a coarse time granularity forecast In this approach, we keep the central period unperturbed, but restrict the times at which vessels may make port-to-port voyages in the forecast period. In all of our computational testing, we use a time grain of 2, which means that vessels may only leave in even numbered time periods. Increasing the time grain above 2 in these instances always resulted in a performance degradation. This reduces the number of travel arcs, but not the number of demurrage arcs.

Rolling horizon heuristic with interspersed calls to K-opt Since solving a 90-period instance to provable optimality can be challenging in its own right, this hybrid is meant to clean up flaws in the solution that have accrued in the process of running a rolling horizon heuristic. Consequently, in this hybrid, a K-opt algorithm is called immediately after $T/2$ periods have been fixed and again after all T periods have been fixed.

Rolling horizon heuristic with restricted port-vessel class pairs This hybrid applies a rolling horizon heuristic over the entire planning horizon after restricting port-vessel class pairs (i.e., after solving MILP (3.6)). Thus, if a port-vessel class pair is deemed incompatible by MILP (3.6), this pair is excluded for the entire solution process thereafter. The hope is that solving each portion of the rolling horizon heuristic is simplified due to these port-vessel class restrictions and that this simplification will result in superior solutions given a time limit.

Algorithm 3 Rolling horizon heuristic with restricted port-vessel class pairs

1: Solve MILP (3.6) to obtain port-vessel class incompatibility pairs $\hat{z}_{j,vc}$
2: Set $x_a^{vc} = 0\ \forall vc \in \mathcal{VC}, a = ((j_1, t_1), (j_2, t_2)) \in \mathcal{A}^{vc}$ s.t. $\hat{z}_{j_1,vc} = 1$ or $\hat{z}_{j_2,vc} = 1$
3: Call Algorithm 1

K-opt with restricted port-vessel class pairs This hybrid applies K-opt over the entire planning horizon after restricting port-vessel class pairs (by solving MILP (3.6)).

Algorithm 4 K-opt with restricted port-vessel class pairs

1: Solve MILP (3.6) to obtain port-vessel class incompatibility pairs $\hat{z}_{j,vc}$
2: Set $x_a^{vc} = 0 \; \forall vc \in \mathcal{VC}, a = ((j_1, t_1), (j_2, t_2)) \in \mathcal{A}^{vc}$ s.t. $\hat{z}_{j_1,vc} = 1$ or $\hat{z}_{j_2,vc} = 1$
3: Call Algorithm 2

3.4 Computational Experiments

Our computational experiments are performed on the Group 2 instances of the publicly available maritime inventory routing problem library (MIRPLib), available at `mirplib.scl.gatech.edu` and presented in Papageorgiou et al. [36]. Since its creation in 2012, the MIRPLib website has received over 75,000 visits and has gained the attention of all of the major MILP solvers, including CPLEX, Express, Gurobi, Mosek, and SCIP. This is important as two of the major goals for the library were: (1) to present benchmark instances for a particular class of MIRPs; (2) to provide the MILP community with a set of optimization problem instances from the maritime transportation domain. Indeed, we show that the MILP solvers have gained some ground as they are now able to find feasible solutions to several instances faster than before. Nevertheless, for large instances, matheuristics are far superior. Moreover, the progress in MILP solvers makes it possible to solve larger neighborhoods within the local search framework.

For consistency, we use the same convention for naming instances proposed in [36], which is based on the number of loading and discharging ports/regions, the number of vessel classes, and the number of vessels. This convention is easily understood with an example. Consider an instance named `LR1_DR04_VC03_V15a`. `LR1` means that there is one loading region/port. `DR04` means that there are four discharging regions/ports. `VC03` means that there are three vessel classes. `V15` means that there are a total of 15 vessels (with at least one vessel belonging to each vessel class). Finally, the letter ‘a’ indicates that the production and consumption rates at all ports are constant over all time periods; the letter ‘b’ indicates that these rates may vary over time.

Table 3.1 provides information and statistics for each of the 72 instances. There are 24 Group 2 base instances that have been formulated with three different planning horizons (120, 180, and 360 time periods) yielding a total of 72 instances. The columns are:

- Column 1 (Horizon): Number of periods in the planning horizon
- Column 2 (Instance name)
- Column 3 (Difficulty): Easy (E), Medium (M), or Hard (H). An instance is declared Hard (H) if no commercial MILP solver (using default settings) can close more than 10% of the gap (defined in (3.7)) in 1800 s; Easy (E) if at least one commercial MILP solver can close more than 90% of the gap in 1800 s using default CPLEX; and Medium (M) otherwise.

Table 3.1 Statistics and best known results for Group 2 MIRPLib instances (Bold font indicates a new best objective function value or lower bound)

Horizon	Instance name	Difficulty	Original model			Presolved model			Statistics			
			Rows	Cols	Int Cols	Rows	Cols	Int Cols	Objval	Bound	Rgap	Agap
120	LR1_DR02_VC01_V6a	E	1081	1855	1134	1023	1839	1118	33809	33809	0.00	0
120	LR1_DR02_VC02_V6a	E	1441	2979	2258	1389	2909	2188	74982	74982	0.00	0
120	LR1_DR02_VC03_V7a	E	1801	4082	3361	1697	3867	3146	40446	**39318**	2.79	1128
120	LR1_DR02_VC03_V8a	E	1801	4004	3283	1713	3854	3133	43721	43717	0.01	4
120	LR1_DR02_VC04_V8a	E	2161	5216	4495	2047	4959	4238	41657	41277	0.91	380
120	LR1_DR02_VC05_V8a	E	2521	6316	5595	2308	5791	5070	36659	**36088**	1.56	571
120	LR1_DR03_VC03_V10b	M	2401	5643	4682	2297	5493	4532	92941	**83645**	10.00	9296
120	LR1_DR03_VC03_V13b	E	2401	5665	4704	2299	5515	4554	124921	118706	4.98	6215
120	LR1_DR03_VC03_V16a	E	2401	5372	4411	2273	5169	4208	82837	**71635**	13.52	11202
120	LR1_DR04_VC03_V15a	E	3001	7152	5951	2860	6940	5739	**73312**	**72108**	1.64	1204
120	LR1_DR04_VC03_V15b	E	3001	7188	5987	2849	6966	5765	117812	**102148**	13.30	15664
120	LR1_DR04_VC05_V17a	E	4201	11110	9909	3908	10538	9337	**72876**	**71572**	1.79	1304
120	LR1_DR04_VC05_V17b	E	4201	11123	9922	3919	10604	9403	**105766**	**84970**	19.66	20796
120	LR1_DR05_VC05_V25a	M	5041	13412	11971	4665	12673	11232	105328	102755	2.44	2573
120	LR1_DR05_VC05_V25b	E	5041	13368	11927	4707	12772	11331	137107	**125966**	8.13	11141
120	LR1_DR08_VC05_V38a	M	7561	20621	18460	7015	19616	17455	166615	**158499**	4.87	8116
120	LR1_DR08_VC05_V40a	E	7561	20759	18598	7074	19919	17758	**178593**	**171418**	4.02	7175

(continued)

Table 3.1 (continued)

Horizon	Instance name	Difficulty	Original model			Presolved model			Statistics			
			Rows	Cols	Int Cols	Rows	Cols	Int Cols	Objval	Bound	Rgap	Agap
120	LR1_DR08_VC05_V40b	M	7561	20724	18563	7055	19848	17687	200746	**188843**	5.93	11903
120	LR1_DR08_VC10_V40a	M	12961	39298	37137	11729	36678	34517	185538	**180353**	2.79	5185
120	LR1_DR08_VC10_V40b	M	12961	39229	37068	11868	37173	35012	206315	**195690**	5.15	10626
120	LR1_DR12_VC05_V70a	H	10921	30166	27045	10169	28847	25726	**278647**	**267766**	3.90	10881
120	LR1_DR12_VC05_V70b	M	10921	30138	27017	10226	28907	25786	308555	**294441**	4.57	14114
120	LR1_DR12_VC10_V70a	H	18721	57811	54690	17162	54871	51750	**283154**	**274121**	3.19	9033
120	LR1_DR12_VC10_V70b	H	18721	57773	54652	17196	54880	51759	295126	**285440**	3.28	9686
180	LR1_DR02_VC01_V6a	E	1621	2815	1734	1563	2799	1718	52167	**52166**	0.00	1
180	LR1_DR02_VC02_V6a	E	2161	4539	3458	2109	4469	3388	129372	**128106**	0.98	1267
180	LR1_DR02_VC03_V7a	M	2701	6242	5161	2597	6027	4946	**60547**	**58790**	2.90	1757
180	LR1_DR02_VC03_V8a	E	2701	6164	5083	2613	6014	4933	**68153**	**66989**	1.71	1164
180	LR1_DR02_VC04_V8a	M	3241	7976	6895	3127	7719	6638	**66017**	**65274**	1.12	743
180	LR1_DR02_VC05_V8a	M	3781	9676	8595	3568	9151	8070	**58619**	**57260**	2.32	1359
180	LR1_DR03_VC03_V10b	M	3601	8643	7202	3497	8493	7052	**125638**	**106712**	15.06	18926
180	LR1_DR03_VC03_V13b	E	3601	8665	7224	3499	8515	7074	165764	**143695**	13.31	22069
180	LR1_DR03_VC03_V16a	M	3601	8372	6931	3473	8169	6728	**143178**	**120556**	15.80	22622
180	LR1_DR04_VC03_V15a	M	4501	10992	9191	4360	10780	8979	**118621**	**116211**	2.03	2410
180	LR1_DR04_VC03_V15b	E	4501	11028	9227	4349	10806	9005	189989	**170432**	10.29	19557

(continued)

Table 3.1 (continued)

Horizon	Instance name	Difficulty	Original model			Presolved model			Statistics			
			Rows	Cols	Int Cols	Rows	Cols	Int Cols	Objval	Bound	Rgap	Agap
180	LR1_DR04_VC05_V17a	M	6301	17110	15309	6008	16538	14737	**117710**	**115530**	1.85	2180
180	LR1_DR04_VC05_V17b	M	6301	17123	15322	6019	16604	14803	**159168**	**130290**	18.14	28879
180	LR1_DR05_VC05_V25a	M	7561	20732	18571	7185	19993	17832	171620	**167766**	2.25	3854
180	LR1_DR05_VC05_V25b	M	7561	20688	18527	7227	20092	17931	205368	**186749**	9.07	18619
180	LR1_DR08_VC05_V38a	H	11341	31901	28660	10795	30896	27655	274244	261036	4.82	13208
180	LR1_DR08_VC05_V40a	H	11341	32039	28798	10854	31199	27958	296760	**284481**	4.14	12280
180	LR1_DR08_VC05_V40b	M	11341	32004	28763	10835	31128	27887	337559	**294265**	12.83	43294
180	LR1_DR08_VC10_V40a	H	19441	60778	57537	18209	58158	54917	304261	**297805**	2.12	6457
180	LR1_DR08_VC10_V40b	H	19441	60709	57468	18348	58653	55412	331775	**304668**	8.17	27107
180	LR1_DR12_VC05_V70a	H	16381	46726	42045	15629	45407	40726	460566	**444823**	3.42	15743
180	LR1_DR12_VC05_V70b	H	16381	46698	42017	15686	45467	40786	491160	**461302**	6.08	29858
180	LR1_DR12_VC10_V70a	H	28081	89371	84690	26522	86431	81750	466975	**454280**	2.72	12695
180	LR1_DR12_VC10_V70b	H	28081	89333	84652	26556	86440	81759	470172	**454552**	3.32	15620
360	LR1_DR02_VC01_V6a	E	3241	5695	3534	3183	5679	3518	108141	**107994**	0.14	147
360	LR1_DR02_VC02_V6a	E	4321	9219	7058	4269	9149	6988	283031	**275347**	2.71	7684
360	LR1_DR02_VC03_V7a	H	5401	12722	10561	5297	12507	10346	**124282**	**120346**	3.17	3936
360	LR1_DR02_VC03_V8a	M	5401	12644	10483	5313	12494	10333	**141166**	**139821**	0.95	1345
360	LR1_DR02_VC04_V8a	M	6481	16256	14095	6367	15999	13838	**138693**	**137449**	0.90	1244

(continued)

Table 3.1 (continued)

Horizon	Instance name	Difficulty	Original model			Presolved model			Statistics			
			Rows	Cols	Int Cols	Rows	Cols	Int Cols	Objval	Bound	Rgap	Agap
360	LR1_DR02_VC05_V8a	H	7561	19756	17595	7348	19231	17070	**122598**	**120986**	1.31	1612
360	LR1_DR03_VC03_V10b	H	7201	17643	14762	7097	17493	14612	**259888**	**212621**	18.19	47267
360	LR1_DR03_VC03_V13b	M	7201	17665	14784	7099	17515	14634	**316441**	**262458**	17.06	53983
360	LR1_DR03_VC03_V16a	M	7201	17372	14491	7073	17169	14288	**327793**	**274753**	16.18	53040
360	LR1_DR04_VC03_V15a	M	9001	22512	18911	8860	22300	18699	**252710**	**248598**	1.63	4112
360	LR1_DR04_VC03_V15b	M	9001	22548	18947	8849	22326	18725	**339308**	**294483**	13.21	44825
360	LR1_DR04_VC05_V17a	H	12601	35110	31509	12308	34538	30937	**251623**	**247816**	1.51	3807
360	LR1_DR04_VC05_V17b	M	12601	35123	31522	12319	34604	31003	**304507**	**251864**	17.29	52643
360	LR1_DR05_VC05_V25a	H	15121	42692	38371	14745	41953	37632	368628	362894	1.56	5734
360	LR1_DR05_VC05_V25b	H	15121	42648	38327	14787	42052	37731	**410053**	**377388**	7.97	32665
360	LR1_DR08_VC05_V38a	H	22681	65741	59260	22135	64736	58255	596969	575369	3.62	21600
360	LR1_DR08_VC05_V40a	H	22681	65879	59398	22194	65039	58558	652380	**624179**	4.32	28201
360	LR1_DR08_VC05_V40b	H	22681	65844	59363	22175	64968	58487	709713	**630407**	11.17	79306
360	LR1_DR08_VC10_V40a	H	38881	125218	118737	37649	122598	116117	663245	649885	2.01	13360
360	LR1_DR08_VC10_V40b	H	38881	125149	118668	37788	123093	116612	724513	651195	10.12	73318
360	LR1_DR12_VC05_V70a	H	32761	96406	87045	32009	95087	85726	1021389	982619	3.80	38770
360	LR1_DR12_VC05_V70b	H	32761	96378	87017	32066	95147	85786	1093013	980410	10.30	112603
360	LR1_DR12_VC10_V70a	H	56161	184051	174690	54602	181111	171750	1024399	1002624	2.13	21775
360	LR1_DR12_VC10_V70b	H	56161	184013	174652	54636	181120	171759	1001541	985670	1.58	15871

- Columns 3–5: Number of constraints (Rows), total number of decision variables (Cols), and number of integer decision variables (Int Cols) in the original model before presolve is called
- Columns 6–8: Same as columns 3–5 except for the presolved model, i.e., the model that is generated after invoking a MILP solver's "presolve" method, a preprocessing routine that reduces the number of variables and constraints, tightens coefficients and variable bounds, and more. The size of the presolve model is almost always more indicative of an instance's difficulty than that of the original model.
- Column 9 (Objval): Best known objective function value ever found (bold if improvement over best known)
- Column 10 (Bound): Best lower bound ever found (bold if improvement over best known)
- Column 11 (Rgap): Best known relative optimality gap $\frac{(z^{\text{best}} - z^{\text{bound}})}{z^{\text{best}}}$
- Column 12 (Agap): Best known absolute optimality gap $(z^{\text{best}} - z^{\text{bound}})$

The main metric that we use to compare solution methods is the average fraction of the relative gap closed over time, where the average is taken over all Easy, Medium, or Hard instances. Here the relative gap is defined as

$$\min\left\{\frac{(z^{\text{method}} - z^{\text{best}})}{z^{\text{best}}}, 1\right\} \tag{3.7}$$

where z^{method} is the objective function value of the method under evaluation and z^{best} is the objective function value of the best known solution ever found. It should be noted that the previous incumbent solutions as reported in [36] were found by warm-starting a local search procedure with the best solution found using the ADP methods described in [34], applying local search for five hours, and then warm-starting Gurobi with this solution and solving for 10 days.

Table 3.2 lists the algorithms compared in this computational study. All matheuristics were coded in AIMMS version 4.13.1.204. Because we set a time limit to search over each neighborhood in our local search implementation, our local search is no longer deterministic due to idiosyncrasies of MILP solvers.

The time limit for solving Model (1) in Step 1 of Algorithm 1 for the rolling horizon heuristic is 120 s, regardless of the number τ^c of central periods.

3.4.1 Benchmarking Experiments with Commercial Solvers

For benchmarking purposes, we first compare the performance of the general purpose MILP solvers CPLEX 12.6.2 and Gurobi 6.5. This comparison is relevant because both solvers have been in possession of these instances since 2012 and have made improvements over the previous versions which were used in Papageorgiou et al. [36]. Two variants were run for each solver. The first directly solves Model (3.3),

Table 3.2 Algorithms compared

Algorithm	Description	Section
cpx	CPLEX 12.6.2 default	–
cpx_mir	CPLEX 12.6.2 with MIR cuts (3.5)	–
cpx_lb	CPLEX 12.6.2 with local branching	–
grb	Gurobi 6.5 default	–
grb_mir	Gurobi 6.5 with MIR cuts (3.5)	–
rhh_τ^c_τ^f	Rolling horizon with τ^c central periods and τ^f forecast periods	Sect. 3.3.1
1opt	Fix all but one vessel class per iteration: $\mathcal{S} = \mathcal{VC}$ in Algorithm 2	Sect. 3.3.2
2opt	Fix all but two vessel classes per iteration: $\mathcal{S}$ = all $\binom{\lvert\mathcal{VC}\rvert}{2}$ pairs in Algorithm 2	Sect. 3.3.2
2opt_cyclic	2opt with many fewer pairs: $\mathcal{S} = \{(vc_1, vc_2), (vc_2, vc_3), \ldots, (vc_{\lvert\mathcal{VC}\rvert}, vc_1)\}$	Sect. 3.3.2
1opt_1daytw	1opt with 1-day time window flexibility search	Sect. 3.3.2
pool_polish	Solution pool polishing	Sect. 3.3.2
rhh_cg_τ^c_τ^f	rhh_τ^c_τ^f with coarse grain forecast	Sect. 3.3.3
rhh_τ^c_τ^f_kopt	Hybrid of rhh_τ^c_τ^f and K-opt	Sect. 3.3.3
rhh_τ^c_τ^f_restrict	Hybrid of rhh_τ^c_τ^f and port-vessel class restrictions	Sect. 3.3.3
kopt_restrict	Hybrid of kopt and port-vessel class restrictions	Sect. 3.3.3

while the second solves Model (3.3) with MIR cuts (3.5) appended to the model for all $t \in \mathcal{T}$ for which the right hand side (3.5) is positive. It must be noted that CPLEX has an advantage when MIR cuts are included as these cuts are added as "User cuts" to a user cut pool, whereas they had to be included in the initial formulation for Gurobi. This distinction is important because when cuts are included in a cut pool, a solver only includes the cuts on an as-needed basis (i.e., when they are violated). Because we could not pass these cuts to Gurobi in a cut pool through the AIMMS-Gurobi interface, Gurobi had to include all of the cuts as structural constraints from the outset resulting in a larger constraint matrix that must be preprocessed and factorized throughout the branch-and-cut algorithm. These benchmarking results with CPLEX and Gurobi were run on a Sandy Bridge dual-socket Linux machine with kernel 2.6.32-279.el6.x86_64 equipped with eight 2.6 GHz cores and 4 GB of RAM per core (32GB of RAM). This machine is more powerful than the machine used to test our matheuristics.

Of the 72 total instances (24 base instances each formulated with a time horizon of 120, 180, and 360 periods), there are 26 Hard instances, 25 Medium instances, and 21 Easy instances. Figure 3.1 further breaks down the number of instances in each difficulty category based on the number of discharging ports and number of time periods in the planning horizon.

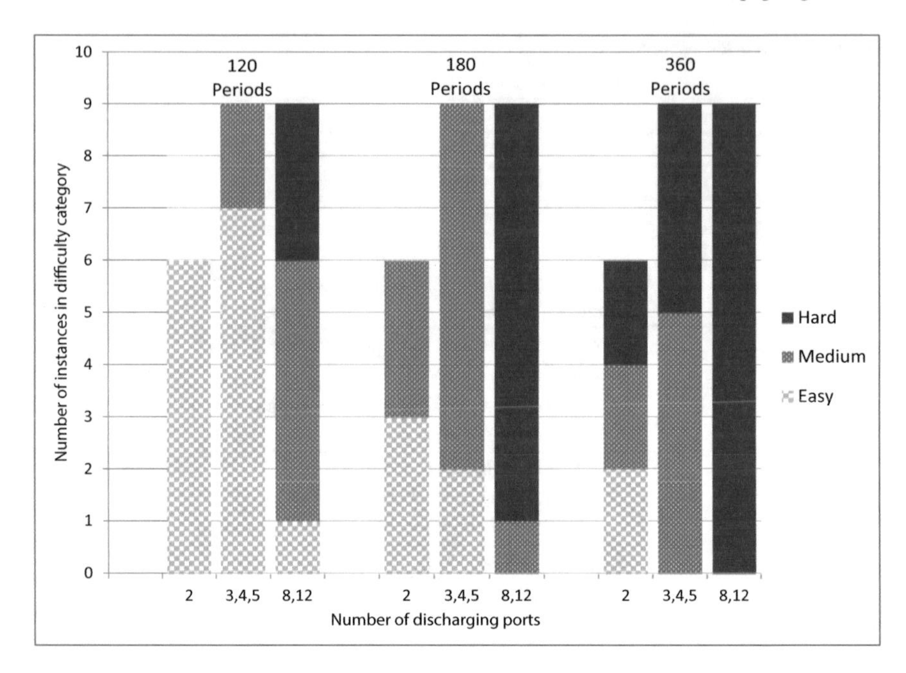

Fig. 3.1 Difficulty categorization of instances evaluated

Figure 3.2 shows profile curves of the average fraction of the gap closed over time broken down by Easy, Medium, and Hard instances. We see that default CPLEX on average outperforms default Gurobi in each difficulty category. As mentioned above, the comparison between both solvers with MIR cuts is not a fair one since it is not an "apples-to-apples" comparison; nevertheless, we report it for completeness. CPLEX with MIR cuts appears to perform better than default CPLEX for Medium and Hard instances. It is important to note that our gap metric (3.7) is different from the traditional gap metric used by solvers. If the latter were used, the benefits of including MIR cuts would be more pronounced as the dual bound improvements are greater. In all subsequent experiments evaluating matheuristics, CPLEX 12.6.2 was used.

3.4.2 Experiments with Construction Heuristics

Figure 3.3 presents performance profiles for several matheuristics and categorizes their performance based on their ability to solve Easy, Medium, and Hard instances. To improve readability, the RHH variants are shown in the top row, K-opt variant in the middle row, and the "best" RHH, K-opt, and commercial solver algorithms are compared in the bottom row. The word "best" is in quotations to emphasize that it is

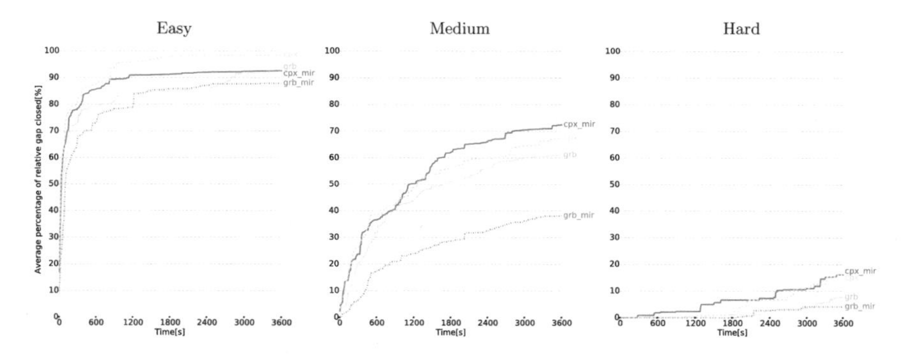

Fig. 3.2 Performance profiles for commercial MILP solvers

based on the performance up to 1800 s. If a shorter time limit were used, the "best" algorithm may change.

Rolling horizon comparison. The top row of Fig. 3.3 compares the performance of the various RHH variants. For all three difficulty levels, RHH variants outperform the best commercial solver within an 1800 s time limit. On Hard instances, `rhh_60_30` performs the best up to 1200 s, after which the two `rhh_30_60` variants begin to take over. Surprisingly, using a coarse grain forecast with `rhh_60_30` on Hard instances is more of a bane than a boon. Using a 90-period forecast was never beneficial on Medium and Hard instances.

***K*-opt comparison**. The middle row of Fig. 3.3 compares the performance of the various K-opt approaches. Although this class of matheuristics was introduced as improvement heuristics in Sect. 3.3.2, they can just as well be applied in an iterative manner to generate high quality feasible solutions. In all of our experiments using a K-opt variant as a construction heuristic, we first solved an entire instance for 30 s before starting Algorithm 2. The results are somewhat mixed. It is not clear if using port-vessel class restrictions provides improvement. For Easy instances, restrictions seem to help, but they are not necessary since CPLEX's performance is satisfactory. For Hard instances, they seem to help for 1-opt as `1opt_restrict` outperforms `1opt`, but the reverse is true when comparing `2opt_restrict` and `2opt`.

Best of the best comparison. The bottom row of Fig. 3.3 reveals, not surprisingly, that the best RHH is superior to the best K-opt variant when used as a construction heuristic. Not shown in the bottom row of Fig. 3.3 is that on Hard instances, in 900 s, `rhh_60_30` performs better than `2opt`, which in turn does better than `rhh_30_60`. This is because `rhh_60_30` is taking larger steps when rolling forward compared to `rhh_30_60`. It should also be reiterated that "order matters" and that perhaps with a more intelligent ordering of vessel class pairs, `2opt` could outperform `rhh_60_30`.

Finally, it is worth calling attention to the fact that on Hard instances, the worst performing RHH and K-opt variant are far superior to all commercial solvers. We will revisit Hard instances in Sect. 3.4.4.

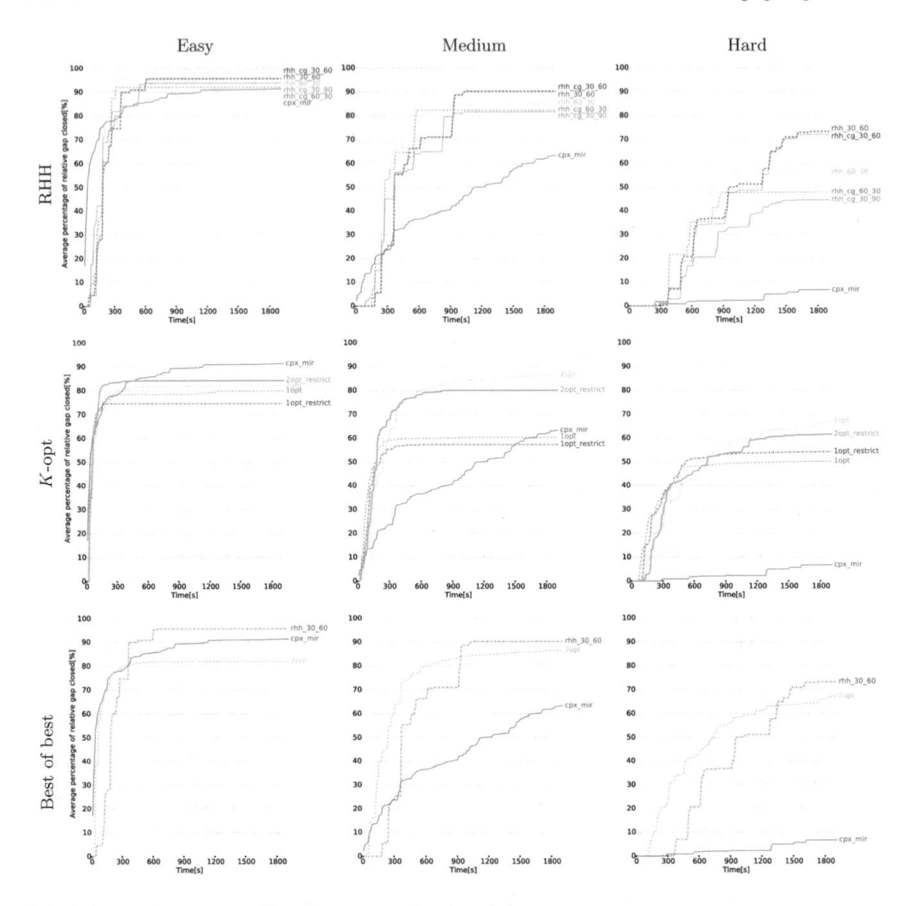

Fig. 3.3 Performance profiles for construction heuristics

3.4.3 Experiments with Improvement Heuristics

In this section, we attempt to evaluate the performance of the improvement heuristics proposed in Sect. 3.3.2. To do so, during the course of our extensive computations, many feasible solutions were generated for each instance. We collected 917 feasible solutions and partitioned them into two groups, which we refer to as buckets: those whose objective function value is less than two times the best known objective function value for that instance (Bucket 1); and those whose objective function is between two and three times the best known objective function (Bucket 2). In other words, Bucket 1 contains "good" solutions and Bucket 2 contains "bad" solutions. For each of the 917 feasible solutions, several improvement heuristics were tested. The percentage improvement presented in this section is given by the ratio $(B - A)/B$, where B is the objective function value of the solution before the algorithm is applied,

Table 3.3 Average percentage improvement of improvement heuristics

	Bucket 1			Bucket 2		
Method	Easy	Medium	Hard	Easy	Medium	Hard
2opt_cyclic	**10.9**	**7.9**	7.2	**50.5**	**47.7**	**34.4**
1opt	8.1	6.5	**8.6**	35.6	32.5	33.5
1opt_1daytw	9.8	5.3	2.8	42.9	35.5	15.5
cpx_lb	10.6	5.2	2.3	45.8	33.5	12.8
cpx	9.8	1.8	0.0	42.8	26.4	1.8

and A is the objective function value of the solution after the algorithm is applied. All improvement heuristics were tested with a time limit of 5 min.

Table 3.3 presents the performance of the different improvement heuristics for "good" solutions (Bucket 1) and "bad" solutions (Bucket 2). The table shows the average improvement of the different heuristics over all instances. Note that an instance can have multiple solutions. Therefore, the percentage improvement of each instance in each bucket is calculated as the average improvement of all the solutions for that particular instance and bucket. The table presents the comparison of 5 methods: 1opt, 1opt_1daytw, 2opt_cyclic, CPLEX with local branching, and default CPLEX. The first three methods are described in Sect. 3.3.2. The last two methods involve solving the problems with CPLEX using local branching (cpx_lb) and default CPLEX (cpx), both using the "incumbent solution" as an initial integer solution.

The reason for evaluating these three K-opt heuristics is because they all require the same number of iterations (the cardinality of the set $\mathcal{S}$ in Algorithm 2 is the same, $|\mathcal{S}| = |\mathcal{VC}|$) with 1opt searching the smallest neighborhood in each iteration, 1opt_1daytw the next largest neighborhood, and 2opt_cyclic the largest neighborhood of the three. The two CPLEX variants search the largest neighborhood as they solve the original problem using an initial solution to warmstart the solution process. Hence, this experiment attempts to answer the question: Empirically, is it better to search larger neighborhoods, which may not solve to provable optimality, or many smaller neighborhoods?

The first thing to note is that, as expected, all improvement heuristics are more successful at improving solutions in Bucket 2 than in Bucket 1. Moreover, all algorithms are better at improving Easy instances, regardless of the bucket. Of all of the tested improvement heuristics, 2opt_cyclic on average performs the best. The only exception is in the Hard instances from Bucket 1, in which 1opt performs the best on average. It is interesting to note that CPLEX with the local branching option performs better than CPLEX with default settings. However, these two do not perform as well as other improvement heuristics, particularly in Medium and Hard instances.

Table 3.4 presents the number of instances for which a better solution was found after applying the different improvement heuristics. For example, if an instance has

Table 3.4 Number of instances for which the method found better solutions

	Easy	Medium	Hard	Total
2opt_cyclic	9	15	19	43
1opt	7	14	22	43
1opt_1daytw	7	13	15	35
cpx_lb	0	0	0	0
cpx	5	4	0	9
pool_polish	5	10	11	26

Table 3.5 Average percentage improvement of best solution using improvement heuristics

	Easy	Medium	Hard
2opt_cyclic	1.3	1.8	3.0
1opt	0.5	1.5	5.4
1opt_1daytw	1.2	0.8	1.2
cpx_lb	0.0	0.0	0.0
cpx	0.8	0.1	0.0
pool_polish	0.5	0.4	0.3

9 solutions in the pool then the improvement is calculated using the best found solution (after applying the heuristic method to all 9 instances), compared to the best solution from the pool. The only exception is pool_polish, in which the best found solution is obtained by applying the solution pool polishing method presented in Sect. 3.3.2 (i.e. this problem is solved only once using all of the solutions, while the other methods are solved one time for each solution).

Table 3.4 shows that 2opt_cyclic and 1opt can improve the best solution in 43 out of the 72 instances. Of these two, 2opt_cyclic is better at improving Easy and Medium instances, while 1opt is better at improving Hard instances. 1opt_1daytw and pool_polish also perform well by improving the best solution in 35 and 26 of the instances, respectively. However, CPLEX has difficulties improving the best solution found. In particular, default CPLEX can only find a better solution in 9 of the instances, while CPLEX with local branching in none.

Table 3.5 presents the average percentage improvement over the best solution from the pool. The quality improvement of the instance in each basket is similar (in terms of behaviour) to the number of instances for which a better solution was found. In particular, 2opt_cyclic performs the best in Easy and Medium instances and 1opt performs the best in Hard instances. Default CPLEX makes little improvement in Easy and Medium instances, and no improvement in Hard instances.

3.4.4 Experiments with Hybrid Heuristics for Hard Instances Only

In this section, we investigate hybrid matheuristics for finding good feasible solutions to the 26 Hard instances in this data set, almost all of which have eight or more discharging regions and five or more vessel classes. These instances are particularly challenging and require extra attention. Likewise, they are more representative of real-world instances.

Three of the hybrid heuristics outlined in Sect. 3.3.3 combine an RHH with a K-opt procedure or port-vessel class restrictions. Figure 3.4 shows the performance of these hybrids. An asterisk means that the CPLEX parameters "Emphasize feasibility over optimality" and "local branching" were engaged. The first observation to make is that all hybrids are able to close between 86 and 91% of the optimality gap in 3600 s. All three RHH methods with $\tau^c = 60$, i.e., 60 periods in the central period, close at least 82% of the optimality gap by 1800 s, whereas the best RHH method tested in Sect. 3.4.2 was able to close roughly 73% of the optimality gap in the same time limit. These results give an affirmative answer to the question: Is it true that RHH-Kopt hybrids are successful at cleaning up flaws in the solution that have accrued in the process of running a rolling horizon heuristic?

Although the heuristic `rhh_60_30_1opt_restrict*`, which includes port-vessel class restrictions, falls to the bottom of the list after 3600 s, it is actually the best performing algorithm up to 1800 s. This suggests that restrictions can be useful for accelerating time to good feasible solutions.

3.4.5 Recommendations for Obtaining Good Solutions Quickly

We close this section by offering a set of recommendations for identifying good solutions reasonably quickly as a function of instance difficulty, the number of time periods, and user-imposed time limits. In general, solver performance is problem dependent, so our recommendations may not hold for other classes of vehicle routing problems.

First and foremost, when relying on a generic MILP solver within a matheuristic, it is important to determine the limits of the underlying solver. The Difficulty column of Table 3.1 reveals that instances with up to three vessel classes and up to 180 time periods are classified as Easy or Medium difficulty. With this empirical data in hand, we felt confident that a MILP solver could make meaningful progress using a 1- or 2-opt local search (i.e., allowing no more than two vessel classes to be free at a time) or solving an RHH of up to 90 periods.

As far as improvement heuristics are concerned, Sect. 3.4.3 reveals that K-opt heuristics are superior to CPLEX with or without local branching for instances with 180 or more time periods. This suggests that these problem-specific heuristics should

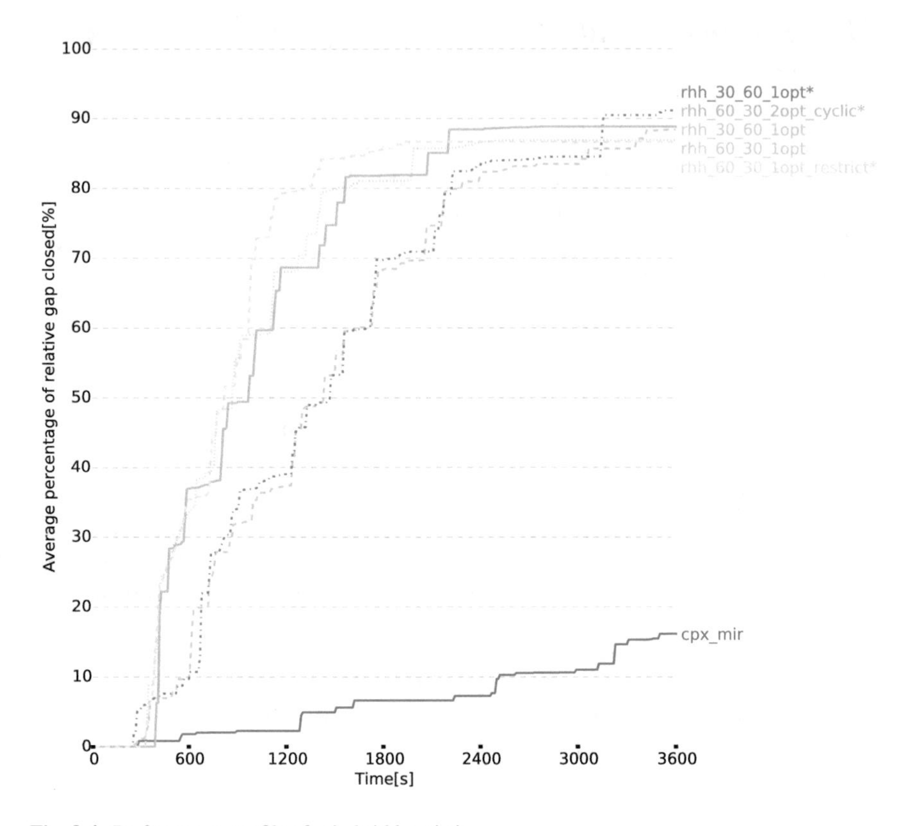

Fig. 3.4 Performance profiles for hybrid heuristics

be used when trying to improve larger instances in one fell swoop. Alternatively, one can break down the problem into smaller chunks and apply a generic MILP solver directly. For example, one could modify the rhh_τ^c_τ^f_kopt algorithm so that K-opt is applied only to the last $R\tau^c$ periods when it is invoked, where R is a positive scalar, instead of to the entire "past."

For Hard instances, and thus most industrial-scale problem instances, arguably the simplest, yet most effective algorithm is coupling a rolling horizon heuristic with a K-opt procedure (rhh_τ^c_τ^f_kopt) as outlined in Sect. 3.3.3. If a relatively long time limit is set, then an RHH with a short central period τ^c (30 in our experiments) allows one to gradually build a solution and make numerous corrections over time. Interspersing calls to a K-opt procedure will further refine suboptimal decisions made in the "past." If a short time limit is imposed, then an RHH with a longer central period τ^c (60 in our experiments) is a wise choice because it rolls forward with larger steps and thus will lead to a complete solution (i.e., all time periods in the planning horizon will be considered) more quickly. If a very short time limit is imposed, a more specialized algorithm may be necessary (e.g., the ADP approach described in [34]).

3.5 Conclusions and Future Research Directions

The matheuristics evaluated in this chapter require a single mathematical program, are simple to implement, and outperform commercial MILP solvers on instances of practical size. Commercial MILP solvers have improved and continue to improve, but heuristics are still needed to solve large MIRP instances for real-world applications for the foreseeable future. The matheuristics presented in this work rely heavily on the performance of MILP solvers to solve smaller, restricted MILPs. Consequently, an improvement in solver performance typically translates into a direct benefit to the matheuristics proposed here. It should also be noted that solvers continue to make progress in parallel implementations and they allow users to search for optimal parameter settings to improve performance on a class of problems. Eventually, these settings may suffice for large industrial applications. Meanwhile, there is on-going research aimed at developing general-purpose parallel large neighborhood search frameworks for finding high quality primal solutions for MILPs (see, e.g., Munguía et al. [32]).

It would be interesting to test matheurisics for continuous-time formulations, which are quite common in ship routing and scheduling problems. These problems typically involve time windows in lieu of inventory balance constraints. Our initial attempts at solving a continuous-time formulation for this problem failed miserably and thus we abandoned the effort. The most likely reason for this poor performance is due to the fact that this class of problems has few operational constraints and no time windows, two ingredients that typically help to reduce the size of a continuous-time formulation. It would also be interesting to have a detailed comparison of matheuristics for operational MIRPs, which typically have more scheduling related constraints.

Acknowledgements We wish to thank two anonymous referees for their feedback, in particular Reviewer 1 whose perceptive comments helped improve the quality of the chapter.

3.6 Nomenclature

Sets are denoted using capital letters in a calligraphic font, such as $\mathcal{T}$ and $\mathcal{V}$. Parameters are typically denoted with capital letters in italic font or with Greek characters. Decision variables always appear as lower case letters.

Indices and Sets

$t \in \mathcal{T}$	set of time periods with $T = \lvert\mathcal{T}\rvert$
$v \in \mathcal{V}$	set of vessels
$vc \in \mathcal{VC}$	set of vessel classes
$j \in \mathcal{J}^P$	set of production, a.k.a. loading, ports
$j \in \mathcal{J}^C$	set of consumption, a.k.a. discharging, ports
$j \in \mathcal{J}$	set of all ports: $\mathcal{J} = \mathcal{J}^P \cup \mathcal{J}^D$
$n \in \mathcal{N}$	set of regular nodes or port-time pairs: $\mathcal{N} = \{n = (j,t) : j \in \mathcal{J}, t \in \mathcal{T}\}$
$n \in \mathcal{N}_{0,T+1}$	set of all nodes (including a source node n_0 and a sink node n_{T+1})
$a \in \mathcal{A}$	set of all arcs
$a \in \mathcal{A}^v$	set of arcs associated with vessel $v \in \mathcal{V}$
$a \in \mathcal{A}^{vc}$	set of arcs associated with vessel class $vc \in \mathcal{VC}$
$a \in \mathcal{FS}_n^{vc}$	forward star associated with node $n = (j,t) \in \mathcal{N}_{s,t}$ and vessel class $vc \in \mathcal{VC}$
$a \in \mathcal{RS}_n^{vc}$	reverse star associated with node $n = (j,t) \in \mathcal{N}_{s,t}$ and vessel class $vc \in \mathcal{VC}$

Data

B_j	number of berths (berth limit) at port $j \in \mathcal{J}$
C_a^{vc}	cost for vessel class vc to traverse arc $a = ((j_1,t_1),(j_2,t_2)) \in \mathcal{A}^{vc}$
$D_{j,t}$	number of units produced/consumed at port $j \in \mathcal{J}$ in period $t \in \mathcal{T}$
Δ_j	an indicator parameter taking value $+1$ if $j \in \mathcal{J}^P$ and -1 if $j \in \mathcal{J}^C$ at port j from a single vessel in a period
P_j	penalty for unmet demand/excess inventory at port $j \in \mathcal{J}$ in time period $t \in \mathcal{T}$
$Q^v (Q^{vc})$	capacity of vessel $v \in \mathcal{V}$ (capacity of a vessel in vessel class vc)
$S_{j,t}^{\min} (S_{j,t}^{\max})$	lower bound (capacity) at port $j \in \mathcal{J}$ in time period $t \in \mathcal{T}$
$s_{j,0}$	initial inventory at port $j \in \mathcal{J}$

Decision Variables

$\alpha_{j,t}$ (continuous) number of units of unmet demand/excess inventory at port $j \in \mathcal{J}$ in time period $t \in \mathcal{T}$
$s_{j,t}$ (continuous) number of units of inventory at port $j \in \mathcal{J}$ available at the *end* of period t
x_a^{vc} (integer) number of vessels in vessel class $vc \in \mathcal{VC}$ using arc $a \in \mathcal{A}^{vc}$.

3.7 Appendix

Lemma 3.1 *Consider the 2-variable mixed-integer linear sets*

$$\mathcal{S}_{\leq} = \{(x,y) \in \mathbb{Z} \times \mathbb{R}_+ : x - y \leq b\} \ .$$

and

$$\mathcal{S}_{\geq} = \{(x,y) \in \mathbb{Z} \times \mathbb{R}_+ : x + y \geq b\} \ .$$

Then, the inequality

$$x - \frac{1}{1-f_0}y \leq \lfloor b \rfloor \tag{3.8}$$

is valid for $\text{conv}(\mathcal{S}_{\leq})$ *where* $f_0 := b - \lfloor b \rfloor$. *The inequality*

$$x + \frac{1}{1-f_0}y \geq \lceil b \rceil \tag{3.9}$$

is valid for $\text{conv}(\mathcal{S}_{\geq})$ *where* $f_0 := \lceil b \rceil - b$.

Proof See [16]. □

Consider a mixed-integer linear set defined by a single constraint:

$$\mathcal{S} = \left\{(\mathbf{x}, \mathbf{y}) \in \mathbb{Z}_+^n \times \mathbb{R}_+^p : \mathbf{a}^\top \mathbf{x} + \mathbf{g}^\top \mathbf{y} \leq \mathbf{b}\right\} .$$

Proposition 3.1 *The* Mixed-Integer Rounding (MIR) *inequality*

$$\sum_{j=1}^{n} \left(\lfloor a_j \rfloor + \frac{(f_j - f_0)^+}{1 - f_0} \right) x_j + \frac{1}{1-f_0} \sum_{j: g_j < 0} g_j y_j \leq \lfloor b \rfloor \tag{3.10}$$

is valid for conv($\mathcal{S}$) *where* $f_0 = b - \lfloor b \rfloor$ *and* $f_j = a_j - \lfloor a_j \rfloor$.[1] *When the single constraint in* $\mathcal{S}$ *is written with a* $\geq$ *sign, the MIR inequality becomes*

$$\sum_{j=1}^{n} \left(\lceil a_j \rceil - \frac{(f_j - f_0)^+}{1 - f_0} \right) x_j + \frac{1}{1-f_0} \sum_{j: g_j > 0} g_j y_j \geq \lceil b \rceil \tag{3.11}$$

where $f_0 = \lceil b \rceil - b$ *and* $f_j = \lceil a_j \rceil - a_j$.

Proof Since the proof of the first case with $\mathcal{S}$ expressed with a $\leq$ sign is given in [16], we prove the second case. Note that $a_j = \lceil a_j \rceil - f_j = \lfloor a_j \rfloor + (1 - f_j)$. Relax the constraint $\sum_{j=1}^{n} a_j x_j + \sum_{j=1}^{p} g_j y_j \geq b$ to

$$\sum_{j: f_j \leq f_0} \lceil a_j \rceil x_j + \sum_{j: f_j > f_0} a_j x_j + \sum_{j: g_j > 0} g_j y_j \geq b\,. \tag{3.12}$$

Here, we have thrown out the continuous variables with non-positive coefficients and we have partitioned the integer variables based on the value of their fractional part f_j being greater than or less than the fractional part f_0 of the right-hand side b. Next, rewrite the left hand side of (3.12) as $w + z \geq b$ where w and z are defined as

[1] Note that the term $\frac{(f_j - f_0)^+}{1 - f_0}$ can only increase the coefficient of x_j, which makes the constraint stronger.

$$w := \sum_{j:f_j \le f_0} \lceil a_j \rceil x_j + \sum_{j:f_j > f_0} \lfloor a_j \rfloor x_j \ , \text{ and}$$
$$z := \sum_{j:g_j > 0} g_j y_j + \sum_{j:f_j > f_0} (1 - f_j) x_j \ .$$

Since $w \in \mathbb{Z}$, $z \in \mathbb{R}_+$, and $w + z \ge b$, we can apply inequality (3.9) from Lemma 3.1. This gives $w + \frac{1}{1-f_0} z \ge \lceil b \rceil$ or

$$\sum_{j:f_j \le f_0} \lceil a_j \rceil x_j + \sum_{j:f_j > f_0} \left(\lfloor a_j \rfloor + \frac{1 - f_j}{1 - f_0} \right) x_j + \frac{1}{1 - f_0} \sum_{j:g_j > 0} g_j y_j \ge \lceil b \rceil \ .$$

Note that $\lfloor a_j \rfloor + \frac{1-f_j}{1-f_0} = \lceil a_j \rceil - \frac{f_j - f_0}{1-f_0}$ for all $j \in \{1, \ldots, n\} : f_j > 0$. Note also that for those integer variables j whose fractional part f_j is greater than the fractional part f_0 of the right-hand side, the relaxed coefficient $\lceil a_j \rceil - \frac{(f_j - f_0)^+}{1-f_0}$ in the MIR cut (3.11) is stronger than for those j with $f_j \le f_0$. □

References

1. Agra, A., Andersson, H., Christiansen, M., Wolsey, L.: A maritime inventory routing problem: discrete time formulations and valid inequalities. Networks **62**(4), 297–314 (2013)
2. Agra, A., Christiansen, M., Delgado, A., Simonetti, L.: Hybrid heuristics for a short sea inventory routing problem. Eur. J. Oper. Res. **236**(3), 924–935 (2014)
3. Al-Ameri, T.A., Shah, N., Papageorgiou, L.G.: Optimization of vendor-managed inventory systems in a rolling horizon framework. Comput. Ind. Eng. **54**(4), 1019–1047 (2008)
4. Al-Khayyal, F., Hwang, S.: Inventory constrained maritime routing and scheduling for multi-commodity liquid bulk, Part I: Applications and model. Eur. J. Oper. Res. **176**(1), 106–130 (2007)
5. Andersson, H., Hoff, A., Christiansen, M., Hasle, G., Løkketangen, A.: Industrial aspects and literature survey: combined inventory management and routing. Comput. Oper. Res. **37**(9), 1515–1536 (2010)
6. Archetti, C., Speranza, M.G.: A survey on matheuristics for routing problems. EURO J. Comput. Optim. **2**(4), 223–246 (2014)
7. Asokan, B.V., Furman, K.C., Goel, V., Shao, Y., Li, G.: Parallel large-neighborhood search techniques for LNG inventory routing. Submitted for publication (2014)
8. Bertazzi, L., Speranza, M.G.: Matheuristics for inventory routing problems. In: Hybrid Algorithms for Service, Computing and Manufacturing Systems: Routing and Scheduling Solutions, p. 488. IGI Global, Hershey (2011)
9. Bixby, R., Rothberg, E.: Progress in computational mixed integer programming—a look back from the other side of the tipping point. Ann. Oper. Res. **149**(1), 37–41 (2007)
10. Boschetti, M.A., Maniezzo, V., Roffilli, M., Röhler, A.B.: Matheuristics: optimization, simulation and control. In: Hybrid Metaheuristics, pp. 171–177. Springer (2009)
11. Brouer, B.D., Desaulniers, G., Pisinger, D.: A matheuristic for the liner shipping network design problem. Transp. Res. Part E: Logist. Transp. Rev. **72**, 42–59 (2014)
12. Campbell, A., Clarke, L., Kleywegt, A., Savelsbergh, M.W.P.: The inventory routing problem. In: Crainic, T.G., Laporte, G. (eds.) Fleet Management and Logistics, pp. 95–113. Kluwer (1998)

13. Christiansen, M., Fagerholt, K., Flatberg, T., Haugen, O., Kloster, O., Lund, E.H.: Maritime inventory routing with multiple products: a case study from the cement industry. Eur. J. Oper. Res. **208**(1), 86–94 (2011)
14. Christiansen, M., Fagerholt, K., Nygreen, B., Ronen, D.: Maritime transportation. In: Barnhart, C., Laporte, G. (eds.) Transportation, Handbooks in Operations Research and Management Science, vol. 14, pp. 189–284. Elsevier (2007)
15. Coelho, L.C., Cordeau, J.F., Laporte, G.: Thirty years of inventory-routing. Transp. Sci. **48**(1), 1–19 (2014)
16. Cornuéjols, G.: Valid inequalities for mixed integer linear programs. Math. Programm. **112**(1), 3–44 (2007)
17. Dauzère-Pérès, S., Nordli, A., Olstad, A., Haugen, K., Koester, U., Myrstad, P.O., Teistklub, G., Reistad, A.: Omya hustadmarmor optimizes its supply chain for delivering calcium carbonate slurry to European paper manufacturers. Interfaces **37**(1), 39–51 (2007)
18. Engineer, F.G., Furman, K.C., Nemhauser, G.L., Savelsbergh, M.W.P., Song, J.H.: A Branch-Price-And-Cut algorithm for single product maritime inventory routing. Oper. Res. **60**(1), 106–122 (2012)
19. Fischetti, M., Lodi, A.: Local branching. Math. Programm. **98**(1–3), 23–47 (2003)
20. Fodstad, M., Uggen, K.T., Rømo, F., Lium, A., Stremersch, G.: LNGScheduler: a rich model for coordinating vessel routing, inventories and trade in the liquefied natural gas supply chain. J. Energy Mark. **3**(4), 31–64 (2010)
21. Furman, K.C., Song, J.H., Kocis, G.R., McDonald, M.K., Warrick, P.H.: Feedstock routing in the ExxonMobil downstream sector. Interfaces **41**(2), 149–163 (2011)
22. Godfrey, G.A., Powell, W.B.: An adaptive dynamic programming algorithm for dynamic fleet management, I: Single period travel times. Transp. Sci. **36**(1), 21–39 (2002)
23. Godfrey, G.A., Powell, W.B.: An adaptive dynamic programming algorithm for dynamic fleet management, II: Multiperiod travel times. Transp. Sci. **36**(1), 40–54 (2002)
24. Goel, V., Furman, K.C., Song, J.H., El-Bakry, A.S.: Large neighborhood search for LNG inventory routing. J. Heurist. **18**(6), 821–848 (2012)
25. Goel, V., Slusky, M., van Hoeve, W.J., Furman, K.C., Shao, Y.: Constraint programming for LNG ship scheduling and inventory management. Eur. J. Oper. Res. **241**(3), 662–673 (2015)
26. Halvorsen-Weare, E., Fagerholt, K.: Routing and scheduling in a liquefied natural gas shipping problem with inventory and berth constraints. Ann. Oper. Res. **203**(1), 167–186 (2013)
27. Hemmati, A., Hvattum, L.M., Christiansen, M., Laporte, G.: An iterative two-phase hybrid matheuristic for a multi-product short sea inventory-routing problem. Eur. J. Oper. Res. **252**(3), 775–788 (2016)
28. Hemmati, A., Stålhane, M., Hvattum, L.M., Andersson, H.: An effective heuristic for solving a combined cargo and inventory routing problem in tramp shipping. Comput. Oper. Res. **64**, 274–282 (2015)
29. Hewitt, M., Nemhauser, G.L., Savelsbergh, M.W.P.: Branch-and-price guided search for integer programs with an application to the multicommodity fixed-charge network flow problem. INFORMS J. Comput. (2012)
30. Hewitt, M., Nemhauser, G.L., Savelsbergh, M.W.P., Song, J.H.: A Branch-and-price guided search approach to maritime inventory routing. Comput. Oper. Res. **40**(5), 1410–1419 (2013)
31. Jiang, Y., Grossmann, I.E.: Alternative mixed-integer linear programming models of a maritime inventory routing problem. Comput. Chem. Eng. **77**, 147–161 (2015)
32. Munguía, L.M., Ahmed, S., Bader, D.A., Nemhauser, G.L., Shao, Y.: Alternating criteria search: a parallel large neighborhood search algorithm for mixed integer programs. Submitted for publication (2016)
33. Mutlu, F., Msakni, M.K., Yildiz, H., Snmez, E., Pokharel, S.: A comprehensive annual delivery program for upstream liquefied natural gas supply chain. Eur. J. Oper. Res. (2015)
34. Papageorgiou, D.J., Cheon, M.S., Nemhauser, G., Sokol, J.: Approximate dynamic programming for a class of long-horizon maritime inventory routing problems. Transp. Sci. **49**(4), 870–885 (2014)

35. Papageorgiou, D.J., Keha, A.B., Nemhauser, G.L., Sokol, J.: Two-stage decomposition algorithms for single product maritime inventory routing. INFORMS J. Comput. **26**(4), 825–847 (2014)
36. Papageorgiou, D.J., Nemhauser, G.L., Sokol, J., Cheon, M.S., Keha, A.B.: MIRPLib—a library of maritime inventory routing problem instances: survey, core model, and benchmark results. Eur. J. Oper. Res. **235**(2), 350–366 (2014)
37. Rakke, J.G., Andersson, H., Christiansen, M., Desaulniers, G.: A new formulation based on customer delivery patterns for a maritime inventory routing problem. Transp. Sci. **49**(2), 384–401 (2014)
38. Rakke, J.G., Stålhane, M., Moe, C.R., Christiansen, M., Andersson, H., Fagerholt, K., Norstad, I.: A rolling horizon heuristic for creating a liquefied natural gas annual delivery program. Transp. Res. Part C: Emerg. Technol. **19**(5), 896–911 (2011)
39. Rothberg, E.: An evolutionary algorithm for polishing mixed integer programming solutions. INFORMS J. Comput. **19**(4), 534–541 (2007)
40. Savelsbergh, M.W.P., Song, J.: An optimization algorithm for the inventory routing problem with continuous moves. Comput. Oper. Res. **35**(7), 2266–2282 (2008)
41. Shao, Y., Furman, K.C., Goel, V., Hoda, S.: A hybrid heuristic strategy for liquefied natural gas inventory routing. Transp. Res. Part C: Emerg. Technol. **53**, 151–171 (2015)
42. Song, J.H., Furman, K.C.: A maritime inventory routing problem: practical approach. Comput. Oper. Res. **40**(3), 657–665 (2013)
43. Stålhane, M., Rakke, J.G., Moe, C.R., Andersson, H., Christiansen, M., Fagerholt, K.: A construction and improvement heuristic for a liquefied natural gas inventory routing problem. Comput. Ind. Eng. **62**(1), 245–255 (2012)
44. Topaloglu, H.: A parallelizable dynamic fleet management model with random travel times. Eur. J. Oper. Res. **175**(2), 782–805 (2006)
45. Topaloglu, H., Powell, W.B.: Dynamic-programming approximations for stochastic time-staged integer multicommodity-flow problems. INFORMS J. Comput. **18**(1), 31–42 (2006)
46. Toriello, A., Nemhauser, G.L., Savelsbergh, M.W.P.: Decomposing inventory routing problems with approximate value functions. Naval Res. Logist. **57**(8), 718–727 (2010)
47. Uggen, K., Fodstad, M., Nørstebø, V.: Using and extending fix-and-relax to solve maritime inventory routing problems. TOP **21**(2), 355–377 (2013)
48. UNCTAD: Review of Maritime Transport United Nations, New York and Geneva (2015)

Chapter 4
Evolutionary Computation for the Ship Routing Problem

Aphrodite Veneti, Charalampos Konstantopoulos and Grammati Pantziou

Abstract In this chapter, we present evolutionary algorithms for solving the real time ship weather routing problem. The objectives to be minimized are the mean total risk and fuel cost incurred along the obtained route while considering the time-varying sea and weather conditions and also a constraint on the total passage time of the route. In addition, for achieving a high safety level the proposed approaches should return only solutions compliant with the guidelines of the International Maritime Organization (IMO). Two different well-known genetic algorithms, namely SPEA2 and NSGA-II are applied to the ship routing problem and a comparative performance evaluation of the two algorithms is performed. The proposed approaches are tested on real data and compared with an exact algorithm which solves the same problem.

Keywords Multi-criteria optimization · Label setting algorithm · Time dependent networks · Resource-constrained shortest path

4.1 Introduction

This chapter presents genetic algorithms for the point-to-point ship weather routing problem which seeks for an optimal route of a ship from a departure to a destination port given a constraint on the total travel time. An optimal route reaches the destination port at minimum fuel consumption and maximum safety taking into account technical and operational restrictions. The problem can be formulated as

A. Veneti (✉) · C. Konstantopoulos
Department of Informatics, University of Piraeus, Piraeus, Greece
e-mail: aveneti@webmail.unipi.gr

C. Konstantopoulos
e-mail: konstant@unipi.gr

G. Pantziou
Department of Informatics, Technological Educational Institution of Athens, Athens, Greece
e-mail: pantziou@teiath.gr

C. Konstantopoulos and G. Pantziou (eds.), *Modeling, Computing and Data Handling Methodologies for Maritime Transportation*, Intelligent Systems Reference Library 131, DOI 10.1007/978-3-319-61801-2_4

a bi-objective, non-linear optimization problem with constraints where the optimal solution should be found between conflicting objectives. Note that the main factor that may determine the optimal route and actually makes ship routing a problem difficult to solve, is the weather conditions. In general, regardless of the particular objectives of the multi-objective ship routing problem, its basic parameters are strongly affected by the current weather and sea conditions. Consequently, the problem of finding an optimal ship route can be considered as a time-dependent shortest path problem. Besides weather variations, moving obstacles with known or unknown trajectories, such as other vessels, as well as marine protected populations or restricted areas are additional factors that make the problem even more dynamic and difficult to solve.

Since solving the ship routing problem takes into account weather forecasts and sea condition, appropriate modelling of the ship response to weather and sea conditions according to the specific characteristics of the ship, is a crucial factor for finding optimal solutions. Note that the weather and sea conditions largely affect all optimization criteria (passage time, fuel consumption and safety) and therefore, an effective solving approach for the ship weather routing problem will necessarily determine the optimal ship power settings and heading control given the particular sea and weather conditions.

A large number of exact as well as heuristic approaches have been proposed in the literature for solving the ship routing problem. Exact algorithms [1, 3, 4, 11, 22, 23, 25, 27, 28, 31–33, 35] derive an optimal solution of the problem, however, at the cost of increased computation time. In contrast, a heuristic approach [10, 15, 20, 24, 29, 34] seeks for a feasible solution only within a subspace of the solution space, hopefully close to the optimal solution, and therefore, the execution time is generally much lower than that of an exact approach. Since, long-term weather forecasts are available, calculus of variations methodology [5] has also proven to be successful for both coastal navigation and trans-oceanic seafaring. Another old but popular method for optimizing ship route is the isochrone method [12].

Although many multi-objective optimization methods exist, evolutionary multi-objective optimization, which applies evolutionary computation to multi-objective optimization, has attracted a great deal of attention. Specifically, there are many different kinds of Multi-Objective Genetic Algorithms (MOGAs), but the common goal is to obtain a Pareto-optimal set that indicates a trade-off relationship. Such studies are common because one characteristic of MOGAs is the multipoint search where multiple Pareto-optimal sets can be obtained by an application of a single search only. For example, in MOGAs, a population of solutions is maintained and used as a basis for search. As the population contains a number of individual solutions, the search makes use of multiple points in the search space. Although, many MOGAs have been proposed in the literature, SPEA2 [40] by Zitzler et al. and NSGA-II [9] by Deb et al. provide excellent results as compared with other MOGA approaches.

The focus of this chapter is on the study of MOGA techniques for the ship routing problem, using as a case study the area of the Aegean Sea. Apart from finding optimal routes with respect to economical objectives, we are also interested in reducing the possibility of an accident in the sensitive area of the Aegean Sea. The derived routes are optimized with respect to two conflicting objectives, the total fuel consumption and the mean risk of the route while taking into account the prevailing weather conditions. In this study, SPEA2 and NSGA-II were both applied to the optimal ship routing problem. Moreover, the performance of these algorithms were compared with that of an exact algorithm [35] for the same problem in order to assess the ability of the two MOGAs in finding the whole Pareto set. It is also worth mentioning that diversity of the obtained solutions and early convergence in MOGAs is achieved by employing the appropriate evolutionary operations and through the appropriate selection of the initial population.

The rest of the paper is organized as follows. In Sect. 4.2, we formally define the point-to-point ship routing problem as a special case of the dynamic multi-objective shortest path problem. The related work is presented in Sect. 4.3. Next, in Sect. 4.4, we describe the proposed genetic algorithm. Some critical modelling issues affecting the effectiveness of solving methods for the ship routing problem are presented in Sect. 4.5. The efficiency of the evolutionary computation in this maritime setting is evaluated in a number of experiments in Sect. 4.6, where both SPEA2 and NSGA-II algorithms are compared against the algorithm in [35] which is a forward label setting algorithm that has the best so far performance in the literature for the problem at hand. Finally, some concluding remarks and the directions for future work are presented in Sect. 4.7.

4.2 Problem Definition

In this subsection, the ship routing problem is formally defined as a non linear integer programming problem. The ship routing problem in its most general version is an instance of the dynamic multi-objective shortest path problem. However, in this study, we focus on the most realistic version of the point-to-point ship routing problem, where the ship speed does not change along the route and waiting at nodes is not an option.

Let $G = (V, A)$ be a directed graph, where V and A is the set of nodes ($|V| = n$) and A and the set of arcs ($|A| = m$) respectively. Each arc in $(i, j) \in A$ is associated with three time dependent positive costs $c^1_{ij}(t)$, $c^2_{ij}(t)$ and $c^3_{ij}(t)$ which are the travel time, the fuel consumption and the risk respectively along this arc when departure from node i takes place at time t. The frozen arc model of [26] is followed in this study where the above arc costs are assumed to be constant during the arc traversal. The problem can be defined as that of finding a shortest path P between a source $s \in V$ and a destination $d \in V$ with minimum total fuel consumption $FC(P)$ and risk $R(P)$, when the departure time at the source node is t_{start} and the voyage duration $VD(P)$ does not exceed a certain upper bound T on the maximum duration of the route.

For the problem formulation, we need an upper bound H on the number of arcs comprising a solution path. This is required because when waiting at nodes is forbidden, the optimal solution may contain infinite number of arcs, as a result of infinite loops along the route [26]. In fact, the obtained solution might be a walk and not a path. Since, this infinite number of arcs cannot be formulated with finite number of variables, an additional constraint on the arc number in a solution path should be enforced.

A number of variables are used for the formulation of problem as a non linear integer programming problem. Namely, we define $p_i \in \{0, 1\}$ with $p_i = 1$ when the solution path has exactly i arcs. We also use the variable α_{ij}^r with $\alpha_{ij}^r = 1$ when the arc α_{ij} is the r-th arc starting from s along the route and $\alpha_{ij}^r = 0$, otherwise. Also t_j^r is defined as the arrival time at node j immediately after traversing arc α_{ij} but only when that arc is the r-th arch of the solution path ($\alpha_{ij}^r = 1$). If such an arc does not exist, we assign a large value to t_j^r. In addition, we always set $t_s^0 = t_{start}$.

$$min\ z = (FC(P), R(P)) \tag{4.1}$$

$$FC(P) = \sum_{r=1...H} \sum_{(i,j)\in A} c_{ij}^2(t_i^{r-1})\alpha_{ij}^r \tag{4.2}$$

$$R(P) = \sum_{r=1...H} \sum_{(i,j)\in A} c_{ij}^3(t_i^{r-1})\alpha_{ij}^r \tag{4.3}$$

$$VD(P) = \sum_{r=1...H} \sum_{(i,j)\in A} c_{ij}^1(t_i^{r-1})\alpha_{ij}^r \leq T \tag{4.4}$$

$$\sum_{j\in V-\{s\}} \alpha_{sj}^1 = 1 \tag{4.5}$$

$$\sum_{(i,j)\in A} \alpha_{ij}^r \leq 1,\ r = 2 \dots H \tag{4.6}$$

$$\sum_{j\in V-\{d\}} \alpha_{dj}^r = 0,\ r = 2 \dots H \tag{4.7}$$

$$\sum_{i=1..H} p_i = 1 \tag{4.8}$$

$$\sum_{j\in V-\{d\}} \alpha_{jd}^r = p_r,\ r = 1 \dots H \tag{4.9}$$

$$\sum_{j\in V} \alpha_{ji}^r - \sum_{j\in V} \alpha_{ij}^{r+1} = 0,\ r = 1 \dots H,\ i\ \in V - \{d\} \tag{4.10}$$

$$\begin{aligned} t_j^r &= \textstyle\sum_{i\in\{k|(k,j)\in A\}} \alpha_{ij}^r(t_i^{r-1} + c_{ij}^1(t_i^{r-1})) \\ &+(1 - \textstyle\sum_{i\in\{k|(k,j)\in A\}} \alpha_{ij}^r)M,\ j \in V,\ r = 1 \dots H \end{aligned} \tag{4.11}$$

$$t_s^0 = t_{start} \tag{4.12}$$

$$c_{ij}^q(M) = M' \tag{4.13}$$

$$\alpha_{ij}^r, p_r \in \{0, 1\}, t_i^r \in R^+, i, j \in V, r = 1 \dots H$$

Constraint (4.4) requires that the voyage duration of the solution path must not be longer than the maximum duration T. Constraint (4.5) ensures that the first arc of the solution path is actually an outgoing arc of the node s. Constraint (4.6) makes sure that at most one arc in the graph can be the r-th edge of the solution path. Constraint (4.7) states that the solution path cannot continue beyond the node d. Constraint (4.8) essentially specifies the number of arcs of the solution path. Constraint (4.9) states that the terminal node of the solution path will be the node d. Constraint (4.10) is a flow conservation constraint ensuring that there is only one incoming and one outgoing arc at each node along the solution path and also that these arcs are consecutive in the path. Equation (4.11) estimates the arrival time at node j when this node is the head of the r-th arc of the solution path. In this equation, we assume that $c_{ij}^{q}(M) = M'$ $(q = 1 \cdots 3)$ where M and M' are large numbers.

Each solution z is the pair of the values of the two objective functions for the corresponding path. Also, although the proposed Integer Programming model is non linear, when the time dependent edge costs functions $c_{ij}^{1}(t)$, $c_{ij}^{2}(t)$ and $c_{ij}^{3}(t)$ are linear, the above formulation can be transformed into a linear model [6].

4.3 Related Work

An Evolutionary Algorithm (EA) is a generic population-based optimization approach which employs techniques inspired by the natural evolution. In an EA the whole solution space is searched randomly for finding an optimal solution. A Genetic Algorithm (GA) is a stochastic optimization algorithm which belongs to the larger class of EAs. A GA maintains a population set of possible solutions. In a number of iterations, the population set is transmuted by genetic operators, generating a new offspring at each iteration. In each generation, parent and offspring compete with each other with the evaluation of solutions being based on a properly defined fitness function. A low value of an individual in this function is essentially the criterion for eliminating that individual from the next generation step. As the search scope extends over the whole solution space and since no constraints are imposed on the fitness function, the GA is able to obtain optimal solutions with high probability. However, using an evolutionary approach for solving dynamic optimization problems is not always straightforward. A common technique in the literature for handling these problems is to increase genetic diversity and prevent early convergence.

Tsou in [34] proposes an evolutionary algorithm for obtaining a number of low cost and high safety routes that avoid narrow waters, bad weather conditions, areas with increased piracy threat, foggy areas, fishing areas, congested areas, etc. The proposed approach employs GIS for spatial data management, spatial analysis and geometric computations. The initial candidate route population is automatically generated by GIS. Specifically, three operations are performed, namely, obstacle detection/avoidance, route generation and route simplification. Then, this set of routes is given as input to the evolutionary algorithm for deriving a larger set of initial routes. Then, a tailor-made evolutionary procedure is applied for route elimination, which derives a set of efficient and feasible routes.

In [29] an evolutionary computation approach has been proposed for solving the multi-criteria ship weather routing problem with criteria the voyage time, the voyage risk and the fuel consumption. It is assumed that the risk depends on the wind speed during the ship voyage. The method is based on the Strength Pareto Evolutionary Algorithm (SPEA) designed for solving combinatorial multi-objective optimization problems. Each solution route is specified as a sequence of waypoints from the origin point to the destination point. The initial population consists of a set of basic routes namely, an orthodrome i.e., a 2-dimensional projection of the shortest curve between the origin point and destination point, a loxodrome i.e., a 2-D projection of the curve intersecting the meridians at a constant angle, a time-optimized isochrone route and another time-optimized isochrone route also optimized for fuel consumption. Then, the initial population is increased by applying random mutations of the basic routes. A method is also developed for route ranking based on the decision-maker preferences, which are expressed in terms of linguistic, fuzzy values. The authors have applied their technique to finding routes in the Atlantic Ocean.

Marie et al. in [24] present a GA for computing the Pareto optimal ship routes with respect to voyage time and fuel consumption. The fuel consumption estimation is based on resistances caused by the wind and the waves. The authors propose a new discretization of the search space based on a number of physical parameters, such as the origin and the destination, the maximum speed, the desired sailing time, the distance between the waypoints and the course changes per hour. The initial population is randomly chosen, as a random sequence of grid nodes. It is assumed that the sea and the wind conditions are time-dependent and a linear interpolation technique is employed for time instances lacking weather measurements.

A GA is proposed in [10] for finding real-time optimal ship routes with optimization criteria the estimated arriva timel, the mean total risk and the fuel consumption. Besides ship movements which affect ship stability, ship structural safety is considered for risk assessment. Ship operational restrictions are also taken into account for solving the optimization problem, while ship course and speed are the control variables. The well-known technique NSGA-II [9] was employed for solving the problem. Among others, the initial route population includes the shortest route in terms of the passage time only. In addition, penalty functions are introduced for preventing constraint violation. Thus, through this artificial increase of the objective value, infeasible solutions are avoided. Finally, an interesting finding in this work is that the Gaussian mutation operator and the two-point crossover operator achieve faster solution convergence.

Since the selection of the initial route population greatly affects the quality of the final solution of an evolutionary approach for solving the ship routing problem, Szlapczynska et al. in [30] proposed the modified isochrone method by Hagiwara in [14], for generating this population. The authors modified the isochrone method so as to ensure that the initial population will not contain routes crossing land since this "no land crossing" property of the initial route population reduces the computational time and improves the quality of the solution. The proposed algorithm has linear computational complexity with respect to the number of grid cells but the optimality of the derived solutions largely depends on the land bitmap resolution. The proposed method is suitable for trans-oceanic navigation, and the authors applied their approach for finding the best route from Plymouth to New York.

In [37], the authors propose a multi-objective route planning method for fishing vessels most suitable for coastal navigation. For faithfully predicting the ship performance in the prevailing sea and weather conditions, detailed models are employed for the seakeeping response of the vessel, the added resistance incurred in an irregular sea state as well as the engine of the vessel. Specifically, graphs of the power rate, fuel consumption and exhaust gases of the engine versus the speed rate are taken into account. The objectives to be optimized are the duration, the fuel consumption and the safety of the trip. For the measurement of the safety level of the trip, a number of criteria are considered, specifically, the slamming and the green water probability, the vertical acceleration at bridge as well as the lateral acceleration. For solving the multi-objective problem, the Strength Pareto Evolutionary Algorithm (SPEA2) is used which determines the Pareto frontier of the solution space. Besides, the heading of the ship, the ship speed is also considered as a decision variable which can be properly set for achieving optimality with respect to the three objectives. Also, for expediting the search of the Pareto-optimal solutions, as a first step, the proposed approach determines the single objective optimal solutions with respect to the three objectives and then these solutions are used as initial solutions for the evolutionary algorithm. Finally, the authors propose a ranking method for the obtained solutions so that the solutions that mostly conform to the user priorities are presented to the user.

In [38] a complete on-board ship weather routing system is described where ship responses are modelled for any sea-state condition taking into account the wave directional spreading. The system employs the multi-objective route planning method described also in Vettor et al. [37], to optimize the route between two ports minimizing fuel consumption, time of arrival and risk related to weather conditions. The performance of the ship weather routing approach has been evaluated in two different simulation scenarios: the first refers to the passage of a container ship departing from the Northern Europe and crossing the North Atlantic Ocean towards the south- east US coast, and the second considers a fishing vessel transiting the western Mediterranean Sea from the port of Valencia to a fishing area south of Malta.

Currently, there is a great deal of active research regarding algorithms and their applications in a variety of multiobjective problems. Among the MOGAs reported to date, the NSGA-II algorithm by Deb et al. and SPEA2 by Zitzler et al. have excellent performance. These algorithms include important search mechanisms, such as preservation of good solutions discovered in the search and reduction of the potential Pareto-optimal solutions.

In our study, the routes are optimized with respect to two conflicting objectives, the total fuel consumption and the voyage risk. For maintaining diversity in the obtained routes, SPEA2 and NSGA-II algorithms have been selected as they are the most promising approaches. The basic operators employed in these algorithms are the node-based crossover operator and three different types of mutation operators. In the initial population, we have included the shortest routes in terms of the traveling time as well as the routes optimized only with respect to a single criterion, either the fuel consumption or the voyage risk. We have also taken into account historic data, and we have added to the initial population all the routes which are usually followed by many ships in practice. Our method takes also into account the IMO restrictions. The performance of the two algorithms is evaluated not only in terms of execution time, but we also assess the ability of the algorithms in retrieving the whole Pareto set.

A preliminary version of the algorithms described in this chapter is presented in [36]. More specifically, a prototype version of the evolutionary algorithm based on the NSGA-II algorithm is described. The objectives of the optimized routes are the same, namely the mean total risk and fuel cost. Also, safety is taken into account and restrictions are applied according to the guidelines of the International Maritime Organization (IMO). However, in the preliminary version, the performance of the algorithm has not been thoroughly examined as in this study.

4.4 Customization of Evolutionary Algorithms

In general, meta-heuristic methods do not ensure the return of the whole set of Pareto-optimal solutions. Another problem is that they often return solutions with no sufficient population diversity. However, NSGA-II and SPEA2 algorithm successfully handle these issues by employing carefully designed operators. In the following, we present the basic operations and structures used by NSGA-II and SPEA2 algorithms in this particular maritime setting.

- Route gene coding: Each solution path (chromosome) comprises a number of nodes (genes) of the sea grid, termed also as waypoints. Each route has a different number of nodes. Also, for each node of the route, its longitude and latitude coordinates are stored as well as the arrival time at that node. Also, all routes are associated with the same ship speed which is assumed to be constant throughout the journey.
- Initialization of route population: The convergence rate of an evolutionary algorithm toward optimal solutions heavily depends on the quality and the diversity of the initial population. Oddly enough, many evolutionary approaches for generating obstacle-free routes perform a random generation of the initial population. Specifically, each route of the initial population comprises random waypoints picked from a predefined navigation area. However, as the random waypoints may be situated over land or obstacles, infeasible routes may be derived. Thus, a random initial population may slow down the convergence rate toward an optimal solution and this in turn may increase the execution time of the algorithm or may result in worse solutions.

In our study, the grid points of a route are exclusively selected from the sea area and not from land or obstacle areas. The initial population contains four different routes, specifically, the routes optimized either for the fuel consumption, navigation, safety or the traveled distance. A main concern in the ship routing problem is also the fact that each weather forecast is relevant only for a specific time interval. As a result, each weather update may alter the optimal routes that have been calculated for the initial population.
Next, the initial population is grown by creating random mutations of the above single-objective optimal routes. By taking into account historical data [39], routes that are followed by many vessels in practice are included in the initial population, as well. However, this sort of information may not be available for all possible departure and destination ports.

- Fitness function: The fitness function indicates the competence of an individual from one generation to the next during the evolutionary computation. Specifically, for the problem at hand, the fitness function should reflect the constraints and the objectives of the problem. Thus, the value of the fitness function for each individual is the pair of values (FC, R) where FC is the total fuel consumption and R is the accumulated risk of the path associated with the individual. Furthermore, the fitness function should be designed in such as way that solutions violating problem constraints will have lower fitness value. In particular, since some kinds of manoeuvre in vessel navigation are not allowed for safety reasons, there should be a constraint on the maximum permitted change of course direction. Lastly, solutions with travel time exceeding the maximum total passage time are discarded from the current population rather than assigned a lower fitness value.
- Route selection in NSGA-II: For elitism preservation and diversity during evaluation, two techniques are used, i.e. the non-dominating sorting and the crowded-comparison operator. Specifically, all solutions are categorized into a number of fronts at different levels and each solution is assigned to its respective front based on its dominance relationship. Thus, the solutions which belong to a front are dominated by at-least one of the solutions belonging to the previous level front. The non-domination rank r_i of a solution corresponds to the level of the front which this individual belongs to. The crowding distance d_i of a solution i is the extent of the search space around i which is not occupied by another solution in the population. Thus, NSGA-II achieves population diversity in the set of the non-dominated solutions by using a niching method.[1] With the above definitions, the crowded comparison operator [8] can now be defined. According to this operator, a solution i is the winner of the tournament with another solution j if (a) solution i has better rank ($r_i < r_j$) or (b) solutions i and j have the same rank but solution i has better crowding distance than solution j, that is, $r_i = r_j$ and $d_i > d_j$.
Essentially, each route within a generation is associated with a fitness value which is determined first by the non-domination level of the route and then by the crowding distance of that route. Now, before producing the individuals of the next generation, the algorithm should select which of the existing individuals (those of the current

[1] a method for finding and preserving multiple favorable parts of the solution space.

and the past generations) will be the candidates for producing the offsprings of the next generation. These individuals should have high fitness values and their number should not exceed a certain size. For determining this set, the existing individuals are first sorted by the crowded comparison operator and then the first in this order individuals are placed in this set. By using the crowded comparison operator, faster convergence toward optimal solutions is achieved since the best solutions in every generation are saved from extinction.
Then, using the crowded comparison operator for deciding the winner, a binary tournament selection operator is applied to the members of the set above for selecting the individuals which will generate the offsprings of the next generation. Next, the new individuals are produced by applying cross-over and mutation operations to the winners of the tournaments.
Finally, the evolutionary process terminates when the performance improvement between successive generations is sufficiently small (smaller than a threshold value) or when a certain number of generations has been produced.
- Route selection in SPEA2: The second approach, the SPEA2 algorithm [40] is a new model of a multi-objective genetic algorithm that improves the search performance of SPEA [41]. The algorithm maintains an archive of fixed size which contains high fitness individuals from all the past generations. The fitness function of individuals in the archive differs from that of individuals in the current population. Specifically, the fitness value of an individual belonging to the archive is the normalized number of the individuals of the current population dominated by that particular individual, whereas for an individual in the current population, the fitness value is the sum of two values: (a) the sum of the fitness values of all archive members which dominate that individual and (b) the density value of that individual where density is defined as a decreasing function of distance of that individual to its k-nearest neighbor and k is a parameter commonly set to the square root of total population size.
Then, the non-dominated individuals from both the current population and the archive will form the updated archive for the next generation. If the total number of these individuals is lower than the fixed size of the archive, then the archive is filled with individuals with the highest overall fitness values from both the current archive and population. If the available non-dominated individuals are more than the required size of the archive, a truncation operation is applied where individuals with many other members of the archive in close proximity are iteratively removed until the size of the archive will be reduced to the pre-determined size. This truncation operation achieves a good spread of non-dominated solutions and also maintains the boundary solutions which are essential for attaining that good spread. After the generation of the new archive, a binary tournament selection with replacement is performed and the mating pool is created. By applying cross-over and mutation operations to the members of the mating pool, the new offsprings of the next generation are produced. The evolutionary process ends when a specified number of generations have been created or if another stopping criterion holds.
- Crossover operation: This operation combines two different parts of two randomly chosen routes (chromosomes) from the mating pool. For making sure that this

combination is valid, the crossover operator from [7] is adjusted. Specifically, for two randomly chosen routes, a random point in the first route is selected for splitting that route (Fig. 4.1). If this point is also a point of the other route, the crossover operation is executed or another route is selected for crossover until the mating pool gets empty.

- Mutation operation: In this operation, there are three possibilities of how a route can be altered [34]. In particular, the first alternative is the Create-disturbance operation (Fig. 4.2) where the coordinates of a node of the route are randomly changed within a previously defined range. Another option is the Insertion operation (Fig. 4.3) where a new node is inserted into the route. More precisely, the new node is picked among all nodes which are inside the circle centered at the midpoint of the edge connecting the two adjacent nodes and with radius being half the edge length. The last possibility is the Deletion operation (Fig. 4.4) where a node along the route is randomly chosen for deletion provided that it is not the departure and destination point or a node in areas specified by the user. It is also guaranteed that all routes obtained from the above operations are not going through land, obstacles or other forbidden areas.

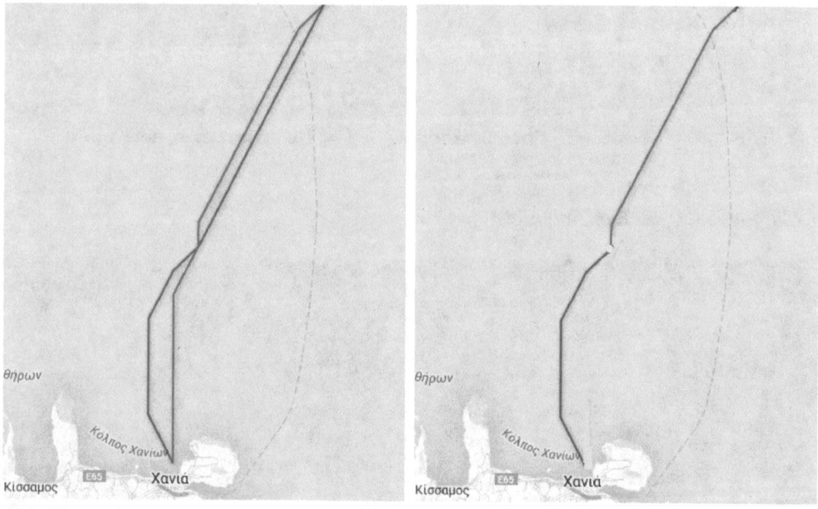

(a) The selected chromosomes routes. (b) Combination of the two randomly selected parts of the chromosomes.

Fig. 4.1 Crossover operation

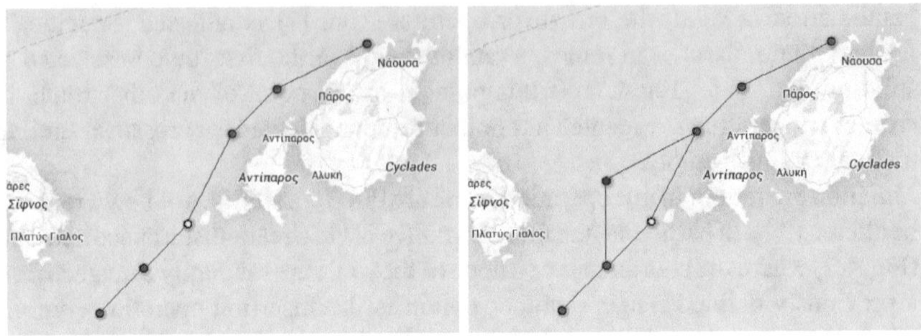

(a) Randomly selected chromosome node. (b) Pertubation of the selected node.

Fig. 4.2 Mutation operation: create-disturbance

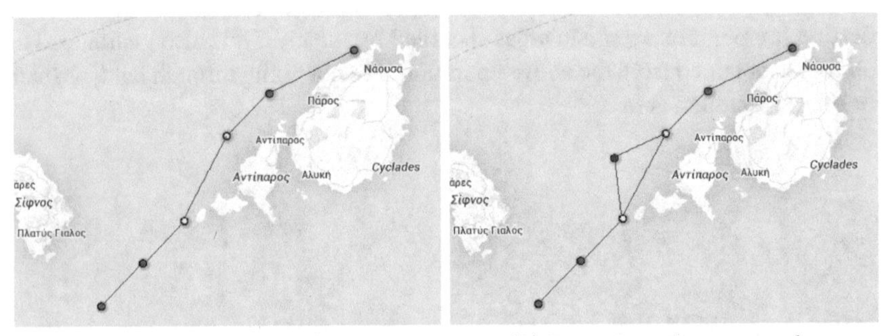

(a) Randomly selected chromosome edge. (b) Insertion of a new node.

Fig. 4.3 Mutation operation: insertion

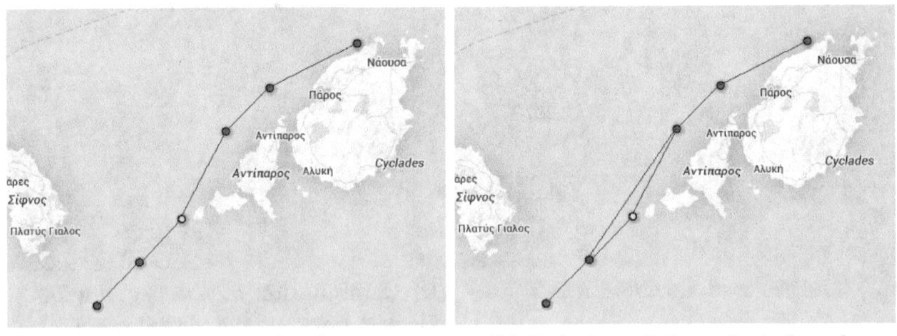

(a) Randomly selected chromosome node. (b) Deletion of the selected node.

Fig. 4.4 Mutation operation: deletion

4.5 Modelization Issues

There is a number of modeling issues that have a critical impact on the effectiveness of ship routing algorithms. Specifically, it is essential to have a ship model which accurately predicts the response of the vessel to the weather conditions and also predicts the effect of these conditions on the real speed of the ship. In addition, an accurate model is needed for the fuel consumption along the ship route since this precise estimation of the consumption is a prerequisite for finding routes with minimum fuel consumption in reality.

We used the Aegean sea as a case study and the grid structure modeling this sea area was developed within the framework of the the AMINESS system [13]. Static and dynamic information was considered, namely geographic and bathymetric data, protected areas, risk estimation [19] as well as predictions for the weather and sea conditions. Moreover, voyage safety was taken into account and dangerous situations were avoided by following the IMO recommendations [18]. In particular, surf-riding and broaching-to are two situations that should be avoided when navigating in non-agreeable weather conditions. Surf-riding and broaching-to arise when $135^\circ < \Theta < 225^\circ$ and $V_R > \dfrac{1.8\sqrt{L}}{cos(180 - \Theta)}$ where Θ, V_R and L are the relative ship-wave angle, the ship speed and the ship length, correspondingly. As the ship speed is constant, the only way to rule out surf-riding and broaching-to is by eliminating the grid edges where these conditions hold.

Parametric rolling motions is another hazardous situation that may arise in bad weather conditions and should also be avoided. It happens when the encounter wave period T_E is almost equal to the natural rolling period of ship T_R or the encounter period T_E is near one half of the ship roll period T_R. The period of encounter T_E is estimated by the following formula:

$$T_E = \frac{3T_W^2}{3T_W + V_R cos(\Theta)} (\text{sec})$$

where T_W is the wave period, Θ is the relative ship-wave angle and V_R is the ship speed (in knots).

For determining the time required for navigating between two points, the actual ship speed should be considered. This speed is often lower than the nominal one because of the added resistance induced by irregular waves and wind during navigation. In this study, we use the model in [23] for predicting the actual ship speed. This is a generic model independent of specific ship features and the speed decrease is a function only the significant wave height H and the ship-wave relative direction Θ. Specifically, the actual ship speed V_R is calculated by the following equation:

$$V_R = V(H, \Theta) = V_0 - f(\Theta) \cdot H^2$$

The values of the coefficient f are listed in Table 4.1.

Table 4.1 Values of the coefficient f

	Sea condition	$f[kn/ft^2]$
$0° \leq \Theta \leq 45°$	following seas	0.0083
$45° < \Theta < 135°$	beam seas	0.0165
$135° \leq \Theta \leq 180°$	head seas	0.0248

If more detailed information about a ship is available, a more analytical model for the vessel response to weather/sea conditions could be employed. Namely, we could use the approximate method by Kwon [21] for estimating the speed loss caused by the added resistance from the wind and rough sea conditions.

The estimation of the fuel consumption rate along a ship route is a very complex issue and a topic of intense research. In practice, the following formula is used: [2, 16, 24, 29, 32]:

$$F = K \cdot P \tag{4.14}$$

where F is the fuel consumption rate measured in kg/h, K is the specific fuel consumption of the ship and P is the engine power in BHP[2] (kW) of the ship. Therefore, the total fuel consumption of the whole route is the product of its passage time PT and the fuel consumption rate F:

$$FC_{total} = F \cdot PT \tag{4.15}$$

4.6 Computational Results

In this section, the two MOGA approaches are compared with a forward label setting algorithm [35], which solves the bi-objective shortest path problem in a time-dependent network, with a maximum travel time constraint and waiting at nodes being forbidden. In our setting, all algorithms search for the Pareto-optimal ship routes between a departure and destination port, optimizing the total fuel consumption and risk, subject to the constraint of the maximum travel time. The departure from the departure port takes place at a fixed time and the nominal ship speed is considered to be constant. Specifically, the test parameters were fixed as follows: speed $= 30$ kn, $K = 200$ g/kWh (specific fuel consumption), engine power $= 4000$ kWh.

All algorithms were implemented in C++ and the tests were carried out on a server with an Intel(R) Xeon(R) E5-2430 v2 at 2.50 GHz processor and 16 GB RAM. Table 4.2 details the parameters of MOGAs in these tests.

[2]Brake HorsePower (BHP) is a measure of the engine power at the output shaft of the engine.

Table 4.2 MOGA parameters

GA parameters	
Population size	100
Terminal generation	100
Crossover rate	0.2
Mutation rate	0.8
Runs	10

Table 4.3 Computational results: execution time

Start	Destination	CPU time (in seconds)		
		NSGA-II	SPEA2	Exact Algorithm
Alonissos	Cythera	18	19	50
Kos	Elafonisi	21	23	69
SW Crete	Chios	20	21	59
Piraeus	Samos	39	42	176
Kalamata	Syros	35	37	158

The execution times of the algorithms for different routes are listed in Table 4.3. For each different route, 100 tests were performed and the average execution time is shown in the third and the fourth column. The execution time of algorithm from [35] is listed in the last column. MOGAs run faster than the algorithm of [35] in all tests. In addition, NSGA-II and SPEA2 have almost the same execution time.

In this study, we used the ratio of non-dominated individuals (RNI) for evaluating the accuracy of the derived solution set and the cover rate for evaluating the breadth of this solution set. Specifically, the ratio of non-dominated individuals is calculated as follows. If SU is the union of the solution sets $S1$ and $S2$ obtained by the two methods, the set of non-dominated solutions in SU is determined and then, the percent of these solutions found by each method is calculated. The higher the value of this ratio for a solution method, the larger the part of the Pareto-optimal front that has been found by this method. Regarding the performance of SPEA2 and NSGA-II, the ratio of non-dominated individuals in SPEA2 was slightly higher (50.9%) than that of NSGA-II (49.1%).

The cover rate of a solution method indicates the diversity of Pareto optimum individuals obtained by the method. This rate is calculated by focusing on a single objective each time. Then, for this particular objective, the distance between the individuals giving the maximum and the minimum value in this objective is calculated and then the interval between the maximum and minimum value is partitioned in subintervals whose number is given as input parameter. Next, the ratio of subintervals containing at least one Pareto optimum individual over the total number of these intervals is estimated. Finally, the same estimation is done for each different objective and then the cover rate is obtained by averaging the ratios above. A high value in the

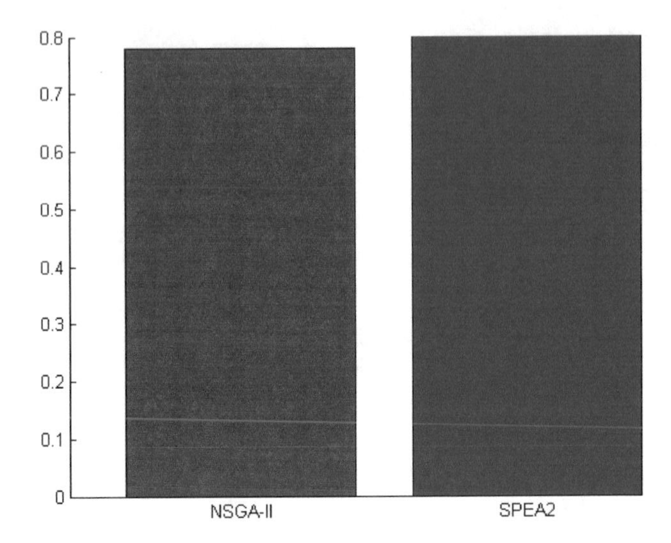

Fig. 4.5 Comparison of NSGA-II and SPEA2 with respect to the cover rate

cover rate clearly means that the Pareto-optimal solutions found are well distributed along the whole Pareto front. From Fig. 4.5, it is evident that the cover rate of the solution set obtained with NSGA-II was slightly lower than that of SPEA2. Also, the results of Fig. 4.6 shows how the cover rate varies with respect to the GA generation.

In principle, the distribution of the solutions could be evaluated by using generalized co-variance value. However, co-variance is not suitable for the solutions that have multiple peaks [17]. In that case, though the co-variance value is high, the diversity of the solutions is very low. For this reason, we opted for the cover rate for evaluating the diversity instead of the co-variance.

For a closer study of the performance of the three algorithms, we also examined the quality of the solution sets obtained for a journey between two specific ports. In Fig. 4.7, we can see the Pareto frontier retrieved by the exact algorithm along with the solution sets derived by the NSGA-II and the SPEA2 algorithm. Solutions of the NSGA-II are denoted with red dots. Blue circles represent the solutions of the SPEA2 algorithm, while solutions of the exact algorithm are denoted with green stars. The exact algorithm returned the whole Pareto set, consisting of 48 routes from Nafplio to Milos (Fig. 4.8a), while the NSGA-II returned 32 routes (Fig. 4.8b) and the SPEA2 retrieved 33 routes (Fig. 4.8c).

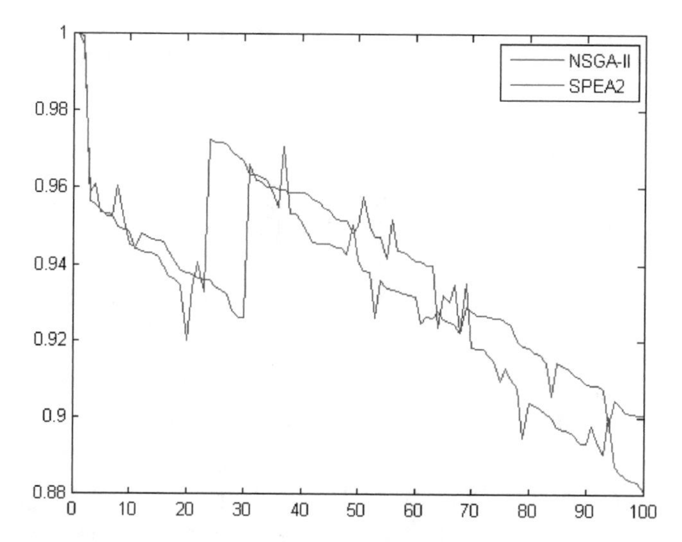

Fig. 4.6 Variation of the cover rate of NSGA-II and SPEA2 versus GA generation

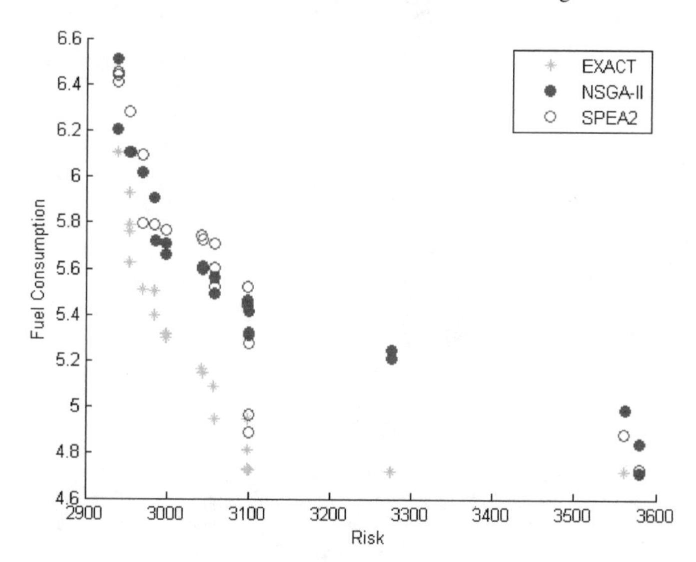

Fig. 4.7 The objective function values of the solutions returned by SPEA2, NSGA-II and the exact algorithm for a route between Nafplio and Milos

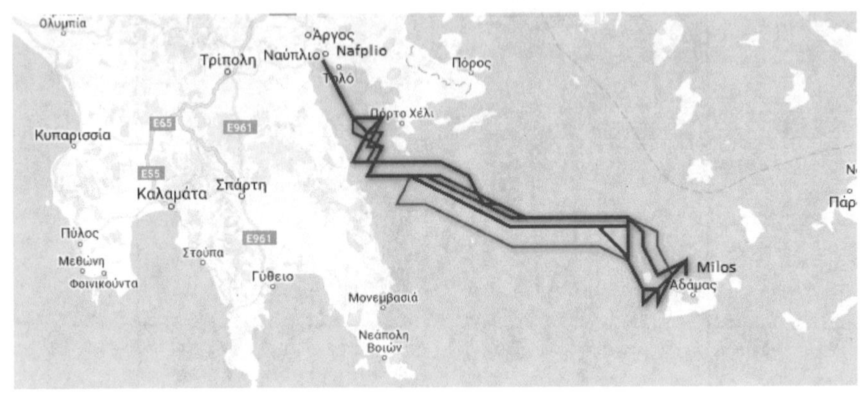

(a) The routes obtained by the exact algorithm.

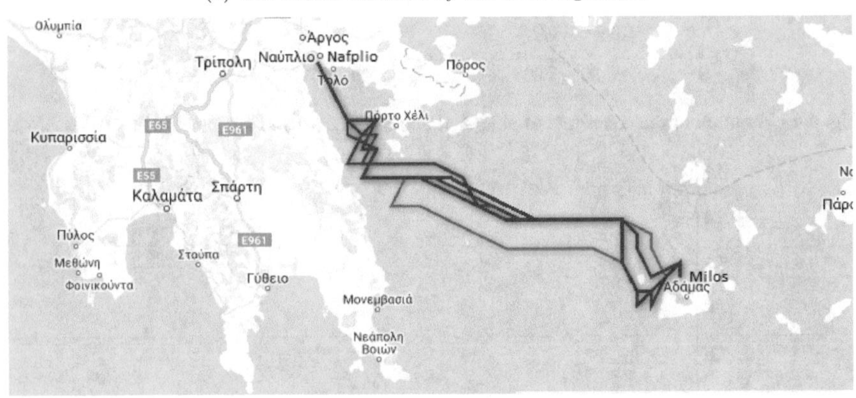

(b) The routes obtained by the NSGA-II algorithm.

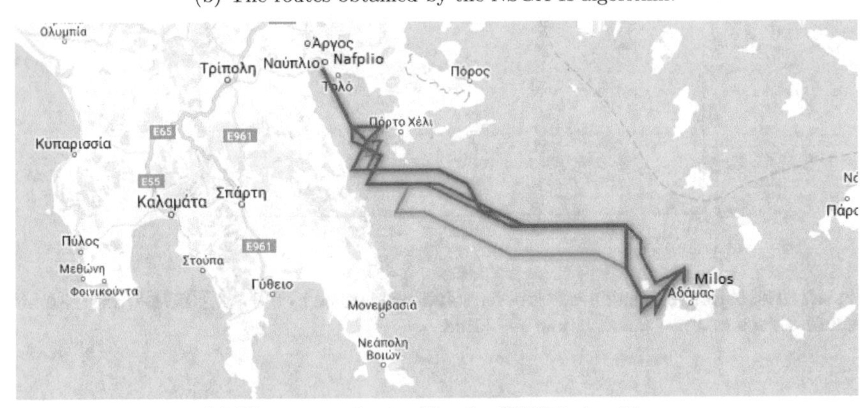

(c) The routes obtained by the SPEA2 algorithm.

Fig. 4.8 Solution sets for the route from Nafplio to Milos

4.7 Conclusion

In this chapter, we studied two evolutionary approaches for solving the time dependent, bi-objective ship routing problem with fixed departure time and a constraint of the maximum voyage duration. All techniques considered use a node based crossover operation and three different kinds of mutation operations. Also, by exploiting historic information, vessel routes which are preferred by many captains are identified and included in the initial population of the two MOGAs. Experimental results show that MOGAs have lower execution time than that of the exact algorithm of [35]. On the downside, the two evolutionary approaches did not manage to retrieve all Pareto-optimal solutions. Concerning the comparison between the two evolutionary algorithms, while NSGA-II slightly outperforms SPEA2 with regard to the execution time, SPEA2 achieves marginally higher accuracy in terms of the ratio of non-dominated individuals and sightly higher cover rate ratio. As a future work, we will study the same problem, allowing the ship speed to change along the route, however not frequently, since frequent speed change is not a common practice for short trips.

Acknowledgements This work was carried out in the framework of the project "AMINESS: Analysis of Marine Information for Environmentally Safe Shipping" which was co-financed by the European Fund for Regional Development and from Greek National funds through the operational programs "Competitiveness and Entrepreneurship" and "Regions in Transition" of the National Strategic Reference Framework—Action: "COOPERATION 2011 Partnerships of Production and Research Institutions in Focused Research and Technology Sectors". The publication of this paper has been partly supported by the University of Piraeus Research Center. Also, in this work the research carried out by the first author was partially funded by Onassis Scholarship Foundation.

References

1. Ari, I., Aksakalli, V., Aydogdu, V., Kum, S.: Optimal ship navigation with safety distance and realistic turn constraints. Eur. J. Oper. Res. **229**(3), 707–717 (2013)
2. Avgouleas, K.: Optimal ship routing. Ph.D. thesis, Massachusetts Institute of Technology (2008)
3. Azaron, A., Kianfar, F.: Dynamic shortest path in stochastic dynamic networks: ship routing problem. Eur. J. Oper. Res. **144**(1), 138–156 (2003)
4. Babel, L., Zimmermann, T.: Planning safe navigation routes through mined waters. Eur. J. Oper. Res. **224**, 99–108 (2014)
5. Bijlsma, S.: A computational method in ship routing using the concept of limited manoeuvrability. J. Navig. **57**(3), 357–369 (2004)
6. Bisschop, J.: AIMMS-Optimization modeling. Paragon Decis. Technol. (2011)
7. Chitra, C., Subbaraj, P.: Multiobjective optimization solution for shortest path routing problem. Int. J. Comput. Inf. Eng. **4**(2), 77–85 (2010)
8. Deb, K.: Multi-objective optimisation using evolutionary algorithms: an introduction. In Multi-objective evolutionary optimisation for product design and manufacturing. Springer, London 3–34 (2011)
9. Deb, K., Pratap, A., Agarwal, S., Meyarivan, T.: A fast and elitist multiobjective genetic algorithm: NSGA-II. IEEE Trans. Evol. Comput. **6**(2), 182–197 (2002)

10. Decò, A., Frangopol, D.M.: Real-time risk of ship structures integrating structural health monitoring data: application to multi-objective optimal ship routing. Ocean Eng. **96**, 312–329 (2015)
11. Dolinskaya, I.S.: Optimal path finding in direction, location, and time dependent environments. Nav. Res. Logist. (NRL) **59**(5), 325–339 (2012)
12. Fang, M.C., Lin, Y.H.: The optimization of ship weather-routing algorithm based on the composite influence of multi-dynamic elements (ii): Optimized routings. Appl. Ocean Res. **50**, 130–140 (2015)
13. Giannakopoulos, T., Vetsikas, I.A., Koromila, I., Karkaletsis, V., Perantonis, S.: Aminess: a platform for environmentally safe shipping. In: Proceedings of the 7th International Conference on PErvasive Technologies Related to Assistive Environments, **45**, 1–45:8 ACM (2014)
14. Hagiwara, H.: Weather routing of(sail-assisted) motor vessels. Ph.D. thesis, Technische Universiteit Delft (1989)
15. Harries S., Heimann J. and Hinnenthal J.: Pareto optimal routing of ships. In: Proceedings of the International Conference on Ship and Shipping Research, Palermo, (2003)
16. Hinnenthal, J., Clauss, G.: Robust pareto-optimum routing of ships utilising deterministic and ensemble weather forecasts. Ships Offshore Struct. **5**(2), 105–114 (2010)
17. Hiroyasu, T., Nakayama, S., Miki, M.: Comparison study of SPEA2+, SPEA2, and NSGA-II in diesel engine emissions and fuel economy problem. In: The 2005 IEEE Congress on Evolutionary Computation, 2005. vol. 1, pp. 236–242. IEEE (2005)
18. IMO, M.: 1/circ. 1228. Revised Guidance to the Master for Avoiding Dangerous Situations in Adverse Weather and Sea Conditions, adopted 11th Jan 2007
19. Koromila, I., Nivolianitou, Z., Giannakopoulos, T.: Bayesian network to predict environmental risk of a possible ship accident. In: Proceedings of the 7th International Conference on PErvasive Technologies Related to Assistive Environments. 44:1–44:5 ACM (2014)
20. Kosmas, O., Vlachos, D.: Simulated annealing for optimal ship routing. Comput. Oper. Res. **39**(3), 576–581 (2012)
21. Kwon, Y.: Speed loss due to added resistance in wind and waves. Nav. Archit. 14–16 (2008)
22. Lo, H.K., McCord, M.R.: Adaptive ship routing through stochastic ocean currents: general formulations and empirical results. Trans. Res. Part A: Policy Pract. **32**(7), 547–561 (1998)
23. Mannarini, G., Coppini, G., Oddo, P., Pinardi, N.: A prototype of ship routing decision support system for an operational oceanographic service. Trans. Nav. Int. J. Mar. Navig. Saf. Sea Trans. **7**(1), 53–59 (2013)
24. Marie, S., Courteille, E., et al.: Multi-objective optimization of motor vessel route. Proc. Int. Symp. Trans. Nav. **9**, 411–418 (2009)
25. Montes, A.A.: Network shortest path application for optimum track ship routing. Technical report, DTIC Document (2005)
26. Orda, A., Rom, R.: Minimum weight paths in time-dependent networks. Networks **21**(3), 295–319 (1991)
27. Padhy, C.P., Sen, D., Bhaskaran, P.K.: Application of wave model for weather routing of ships in the North Indian Ocean. Nat. Hazards **44**(3), 373–385 (2008)
28. Shao, W., Zhou, P., Thong, S.K.: Development of a novel forward dynamic programming method for weather routing. J. Mar. Sci. Technol. **17**(2), 239–251 (2012)
29. Szlapczynska J., Smierzchalski R.: Multicriteria optimisation in weather routing. In: Weintrit A. (eds.) Marine Navigation and Safety of Sea Transport. Taylor & Francis Group, London (2009). ISBN: 978-0-415-80479-0
30. Szlapczynska, J., Smierzchalski, R.: Adopted isochrone method improving ship safety in weather routing with evolutionary approach. Int. J. Reliab. Qual. Saf. Eng. **14**(06), 635–645 (2007)
31. Szlapczynski, R.: A new method of ship routing on raster grids, with turn penalties and collision avoidance. J. Navig. **59**(1), 27–42 (2006)
32. Takashima, K., Mezaoui, B., Shoji, R.: On the fuel saving operation for coastal merchant ships using weather routing. Proc. Int. Symp. Trans. Nav. **9**, 431–436 (2009)
33. Tsatcha, D., Saux, É., Claramunt, C.: A bidirectional path-finding algorithm and data structure for maritime routing. Int. J. Geogr. Inf. Sci. **28**(7), 1355–1377 (2014)

34. Tsou, M.C.: Integration of a geographic information system and evolutionary computation for automatic routing in coastal navigation. J. Nav. **63**(2), 323–341 (2010)
35. Veneti, A., Konstantopoulos, C., Pantziou, G.: Continuous and discrete time label setting algorithms for the time dependent bi-criteria shortest path problem. In: Operations Research and Computing: Algorithms and Software for Analytics, pp. 62–73 (2015)
36. Veneti, A., Konstantopoulos, C., Pantziou, G.: An evolutionary approach to multi-objective ship weather routing. In: 2015 6th International Conference on Information, Intelligence, Systems and Applications (IISA), pp. 1–6. IEEE (2015)
37. Vettor, R., Tadros, M., Ventura, M., Soares, C.G.: Route planning of a fishing vessel in coastal waters with fuel consumption restraint. Marit. Technol. Eng. **3** (2016)
38. Vettor, R., Soares, C.G.: Development of a ship weather routing system. Ocean Eng. **123**, 1–14 (2016)
39. Vodas, M., Pelekis, N., Theodoridis, Y., Ray, C., Karkaletsis, V., Petridis, S., Miliou, A.: Efficient ais data processing for environmentally safe shipping. SPOUDAI-J. Econ. Bus. **63**(3–4), 181–190 (2014)
40. Zitzler, E., Laumanns, M., Thiele, L., et al.: Spea2: Improving the strength pareto evolutionary algorithm. In: Eurogen, pp. 95–100 (2001)
41. Zitzler, E., Thiele, L.: An evolutionary algorithm for multiobjective optimization: the strength pareto approach. Swiss Federal Institute of Technology (ETH), Switzerland **43**, (2008)

Chapter 5
Decision Support Tool Employing Bayesian Risk Framework for Environmentally Safe Shipping

Sotirios Gyftakis, Ioanna Koromila, Theodore Giannakopoulos, Zoe Nivolianitou, Eleni Charou and Stavros Perantonis

Abstract Due to the significant increase of tanker traffic from and to the Black Sea that pass through narrow straits formed by the 1600 Greek islands, the Aegean Sea is characterized by an extremely high marine environmental risk. Therefore it is vital to all socio-economic and environmental sectors to reduce the risk of a ship accident in that area. In this chapter a web tool for environmentally safe shipping is presented. The proposed tool focuses on extracting aggregated statistics using spatial analysis of multilayer information: vessel trajectories, vessel data as well as information regarding environmentally important areas. The decision support system includes preprocessing, clustering of trajectories (based on their spatial similarity) and risk assessment employing probabilistic models (Bayesian network). Applications of the web tool are presented in areas such as marine traffic monitoring in environmentally protected areas, and influence of restricted areas in marine traffic. Results demonstrate that the web tool can provide essential information for maritime policy makers.

Keywords Marine safety · Marine traffic monitoring · Risk analysis · Bayesian network · Dynamic risk model · Vessel trajectories classification and visualization · Spatiotemporal trajectories analysis

---

S. Gyftakis (✉) · T. Giannakopoulos · E. Charou · S. Perantonis
Institute of Informatics and Telecommunications, National Centre for Scientific Research "DEMOKRITOS", Aghia Paraskevi, Greece
e-mail: sotgyft@gmail.com

I. Koromila · Z. Nivolianitou
Institute of Nuclear and Radiological Sciences and Technology, Energy and Safety, National Centre for Scientific Research "DEMOKRITOS", Aghia Paraskevi, Greece

I. Koromila
Ship Dynamics, Stability and Safety Research Group, Department of Naval Architecture and Marine Engineering, National Technical University of Athens, Athens, Greece

C. Konstantopoulos and G. Pantziou (eds.), *Modeling, Computing and Data Handling Methodologies for Maritime Transportation*, Intelligent Systems Reference Library 131, DOI 10.1007/978-3-319-61801-2_5

5.1 Introduction

Over the last years shipping activity has greatly increased since the demand of merchandises between countries in different continents has multiplied. Indeed, cargo transported by the liner shipping industry represents about two thirds of the total global trade value [1]. In addition, the marine environment encompasses an immense diversity of life species. Maritime industry should, therefore, coexist in harmony with marine life. There are several marine areas that comprise significant environmental and ecological sensitive resources. In this chapter, the Greek Seas (Aegean and Ionian) are considered as the geographical scope of the research. Such a region represents an exceptional marine area with not only unique geographical morphology, important environmental and ecological sensitivity, but intense marine traffic as well.

The restriction of the national waters monitoring policy to the marine accident prevention is a significant fact, mainly due to legal limitations when enforcing shipping routes in the Aegean Sea. In fact, although traffic separation schemes (TSS) or similar routing-schemes are active in the Mediterranean Sea (e.g. Straits of Gibraltar, Algeria, Barcelona, Italy, Port Said and the North Adriatic Sea) and the Black Sea (e.g. between the Dardanelles and Istanbul including the Bosporus, ports of Odessa and Sevastopol), currently there is not any TSS in the area of central Aegean Sea where it is particularly essential. Consequently, shipping companies and vessels' masters operating in the waters of the Aegean Sea may neglect risks that would derive from a possible accident, due to unsafe trajectories and local weather conditions. In fact, restricted and shallow waters in conjunction with the particular prevailing weather conditions, such as "meltemia"[1], consist some factors that may contribute to a marine accident.

A possible maritime accident that results in oil spill would negatively affect the marine environment of the Greek Seas. In addition, such an accident would have negative impact on local communities, tourism, fisheries, public health as well as the economy. As accidents involving oil spills are impossible to be completely cleaned up, the Greek Seas ecosystem may suffer long term catastrophic and irreversible changes. Therefore, it is essential to minimize the large environmental risk in order to make the Greek Seas safer. This could be realized through the development of a decision support tool that offers environmental policy recommendation options.

The main capabilities of the proposed decision support system are the following four [7]:

- Integration of information from multiple sources of maritime importance
- Employment of risk assessment
- Incorporation of preprocessed historical maritime data
- Immediate delivery of results for alternate policy options

Briefly explained, the proposed method consists of a trajectory density model combined with a visualization model with two visual cues (color and density). This

[1] Meltemia are very strong, prevailing winds in the area of central Aegean Sea during Summer.

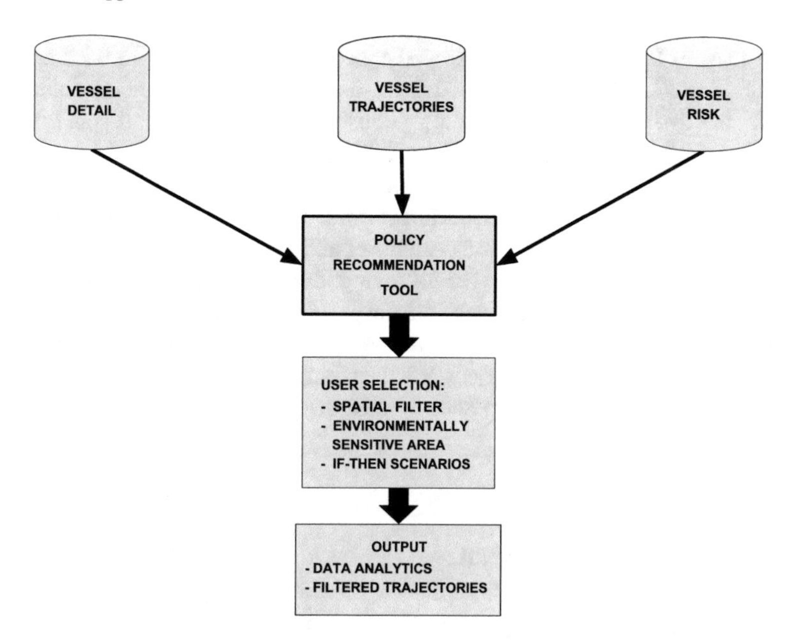

Fig. 5.1 Decision support tool flowchart

methodology is considered proper to be applied in the maritime domain for improving both safety and security.

In Fig. 5.1 the flowchart of the decision support system is shown. The main module is the policy recommendation tool. The input consists of the data providing modules: vessel trajectories, vessel detail, and vessel risk. The user selection module is employed for user interaction with the tool. The selection includes spatial filtering, use of an environmentally sensitive area, and composition of if-then-scenarios. Finally, the output module provides data analytics and stores results for further analysis.

The main purpose of the developed monitoring system is to deliver user-friendly data analytic to policy makers and general users using all types of spatio-temporal marine data, discriminating both long-term vessel trajectories and vessel details. In order to extract long-term statistics and trends, historical marine traffic data are taken into account. Specifically, data extracted from the Automatic Identification System (AIS) are considered. AIS is an automatic tracking system used on ships and by vessel traffic services for identifying and locating vessels.

There are a lot of challenges concerning AIS data visualization and usage for maritime safety. Jiakai et al. [17] propose an AIS data visualization model that includes an index of maritime traffic situation (IMTS). The IMTS index is a weighted function of the ship encounter rate, turn rate and acceleration speed. In a case study, the authors use collected AIS data and argue that their visualization model provides useful information for the maritime traffic decision making. Silveira et al. [28] analyze historical AIS data in order to calculate the expected collision risk for specific areas off the coast of Portugal. Their suggested approach for estimating the risk collision in complex traffic patterns is obtained from the number of collision candidates calculated directly from decoded AIS data and using the concept of collision diameter. The authors suggest that their method provides important insight into the level of maritime risk and improvement of maritime safety. Furthermore, a methodology for computing and visualizing behavioural patterns of vessel trajectories by means of density is presented in the paper of Willems et al. [31].

The structure of this chapter is the following: In Sect. 5.2 the model of vessel risk estimation is developed. Next, in Sect. 5.3 the preprocessing and the clustering of historical trajectories are outlined. The policy recommendation tool is comprehensively discussed in Sect. 5.4 while in Sect. 5.5 decision tool applications are presented. Finally, conclusions are drawn in Sect. 5.6, followed by acknowledgements and the list of references.

5.2 Vessel Risk Estimation

In the context of the present sub module, the marine environmental risk of a vessel is estimated. Aggregated results of that risk module are visualized in the policy recommendation tool in order to provide a view of the developed ship risk distribution.

A risk assessment procedure is applied to estimate the environmental risk. In the field of maritime safety, Formal Safety Assessment guidelines [14] are utilized for assessing risks. A typical FSA study consists of the following basic steps:

Step 1 Hazards identification
Step 2 Risk analysis
Step 3 Risk control options
Step 4 Cost benefit assessment, and
Step 5 Recommendations for decision-making

In order to develop the appropriate environmental risk model, both the potential hazards to vessels are identified and a risk analysis is conducted. Indeed, collision, contact and grounding accidental events are identified as potential hazards to vessels. Such kind of marine accidents might lead to huge oil spills negatively affecting the marine environment. In more detail, a collision event includes ships striking or being struck by another ship, regardless of whether under way, anchored or moored, while a contact event includes ships striking or being struck by an external object, but not another ship or the sea bottom. As a grounding event, a vessel striking the sea bottom, shore or underwater wrecks is considered. Although a lot of efforts, through national and international rules and regulations, have been attempted to prevent these events, accidents continue to occur.

Therefore, collisions, contacts and groundings are taken into consideration in order to conduct the risk analysis step. In line with the FSA guidelines, risk is defined as the probability of an event multiplied by its consequences (see (5.1)).

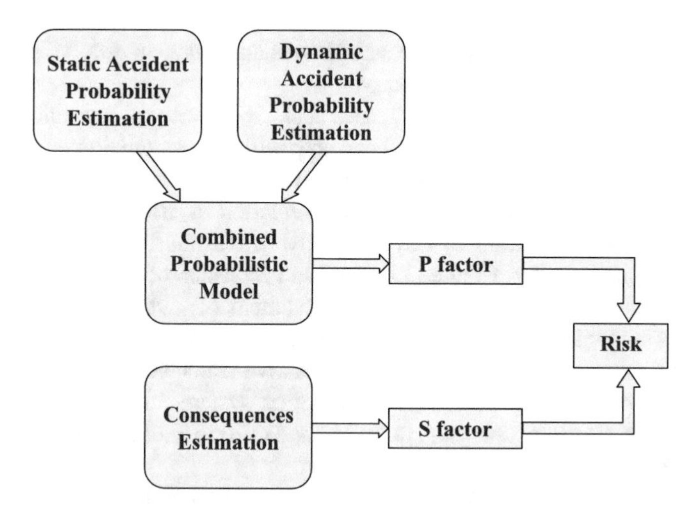

Fig. 5.2 Risk model

$$R = P \times S \tag{5.1}$$

The proposed model combines the results of a marine accident probability (P) with the severity of its consequences (S). The factor P, as a first approximation, is calculated employing the Bayesian network methodology by using a simplified model to predict the probability of an accident given the main characteristics of the vessel; namely the ship type, size, age and flag [20]. A more advanced model taking into account the dynamic factor of the probability is elaborated, along with considerations about the weather, the geographical area and the traffic conditions [21]. As far as the consequences (S) are concerned and considering the environmental scope of this study, attention should be drawn on the estimation of the impact resulting from an oil spill. In the final stage of the model, the two factors (P and S) are integrated and produce the final environmental risk of a possible ship accident in the Aegean Sea. A process flowchart is illustrated in Fig. 5.2. In the following subsections, the components of the proposed model are briefly discussed.

5.2.1 Probability Estimation

The most appropriate probability modelling tool that fits in the present research is the Bayesian network methodology. The Bayesian network is considered to be the most proper methodology among others, namely Fault Tree Analysis (FTA), Event Tree Analysis (ETA), Probabilistic Risk Analysis (PRA), Fuzzy logic, Analytical Hierarchy Process (AHP) and other artificial intelligence methods. There are several researchers involved in the development of Bayesian models in order to enhance the safety level within the maritime industry.

Det Norske Veritas (DNV) [2] and Jensen et al. [16] develop a Bayesian network to calculate the probability of the event that a ship will have a collision or a grounding accident. DNV presents additional quantitative human risk results for these two accident scenarios that relate to failure in navigation of cruise vessels. Hanninen [12] discusses the utilization of Bayesian networks in maritime safety modeling. In addition, Hanninen and Kujala [13] suggest a Bayesian network using port state control inspection data to discover the involvement of a ship in maritime traffic accidents. Another important study is the one of Montewka et al. [24]. This research introduces a systematic, transferable and proactive framework estimating the risk of ship-ship collision for maritime transportation systems. The risk framework is developed with the use of Bayesian networks and utilizes a set of analytical methods for the estimation of the risk model parameters.

In the following subsections a description is presented of the two distinct parts in which the probability estimation is divided; the static and the dynamic model.

5.2.1.1 Static Model

The purpose of the static model is to estimate the accident probability given the static information of the vessel. Figure 5.3 illustrates an overview of that model.

The static Bayesian network takes into consideration four input variables; the type, the size, the age and the flag of the vessel. Each of these variables has a set of values.

Briefly explained, the vessel type could take the values *passenger*, *tanker*, *general cargo*, and *other*. *Passenger* is any vessel carrying more than 12 passengers, *tanker* is any vessel carrying liquid cargo, *general cargo* is any vessel carrying bulk cargoes, containers or vehicles, and *other* is any other vessel, such as research vessel, yacht, tug, and fishing. The variable vessel size can take the values *small*, *medium*, and *large* depending on the weight of the cargo (deadweight) for tankers and general cargoes, and the length for passenger vessels. All other ships are considered as *small*.

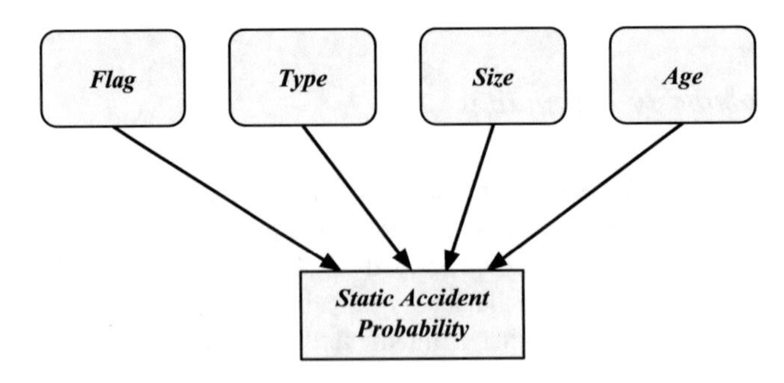

Fig. 5.3 Static model

Table 5.1 States of static model

Node	States			
Accident probability	low	middle	high	
Age	new	middle	old	
Flag	low risk	middle risk	high risk	very high risk
Size	small	middle	large	
Type	passenger	tanker	general cargo	other

As far as the age of the vessel is considered, this variable is split into *new*, *middle* and *old*. A vessel under the age of five years is considered *new*, while a vessel older than twenty-five years is an *old* vessel. Finally, the variable flag of the vessel could take the values *low risk*, *medium risk*, *high risk* and *very high risk*, according to PARIS MOU flag list 2012. The states of each vessel variable are shown in Table 5.1.

Therefore, the output of the static model is the static probability regarding that this vessel might be involved in a collision, contact or grounding event. This output is estimated by computing the probability indicated in (5.2) from the Bayesian Network.

$$P(\mathit{Static\ Accident} = True\ |Type, Size, Age, Flag) \tag{5.2}$$

To estimate the probabilities and therefore, train the Bayesian model, historical data were used in addition to expert judgement. Actually, the Greek Ministry of Mercantile Marine casualty database is the primary data source. In particular, more than 100 accidents have been fully annotated with the respective data vessel records. For each accident both the vessel IMO and MMSI number, type and year of build, the current (if the vessel exists) and previous (at the moment of the accident) vessel name and flag, the type of accident and the location the accident occurred were recorded.

5.2.1.2 Dynamic Model

The dynamic probability corresponds to a more complicated Bayesian network, taking into consideration weather, geographical and traffic conditions. Figure 5.4 illustrates an overview of the proposed dynamic model.

In more detail, the input variables are the prevailing weather conditions and the navigation area, while the output variable is the dynamic probability that the vessel will be involved in a collision, contact or grounding event. The accident probability is estimated through the computation of the result of (5.3).

$$P(\mathit{Dynamic\ Accident} = True\ |\mathit{Weather\ Conditions,\ Navigation\ Area}) \tag{5.3}$$

In the dynamic Bayesian network, both the weather conditions and the navigation area nodes have hidden input nodes that must be taken into consideration. The specific

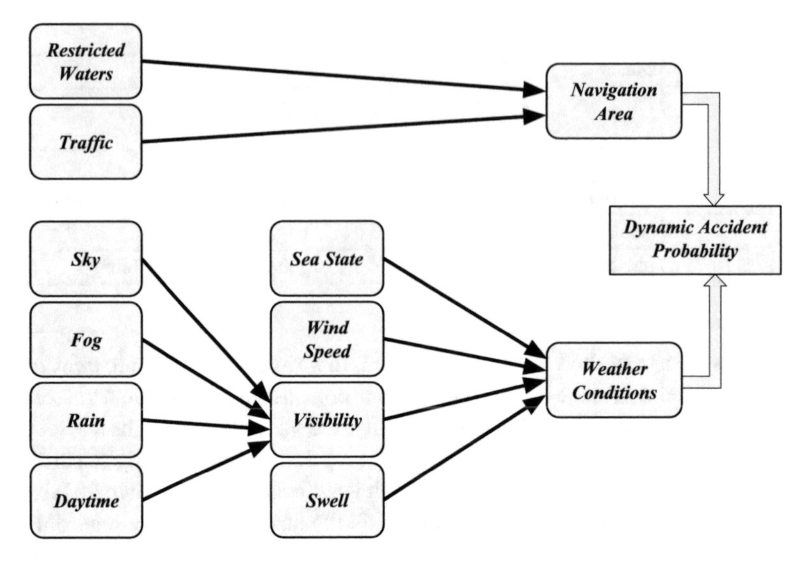

Fig. 5.4 Dynamic model

navigation area characteristics are taken into account in order to define the *low*, *medium* or *high* contribution to marine accidents. Type of waters (e.g. narrow street and shallow waters) and marine traffic density are the navigation areas parent nodes. The prevailing weather conditions are of significant importance for a possible marine accident. *Sea state*, *wind speed*, *swell* and *visibility*, are the appropriate states for that node. Finally, the visibility node is calculated given the values of the *daytime*, *fog*, *sky* and *rain*. Therefore, in order to estimate the dynamic accident probability, the conditional probabilities presented in (5.4), (5.5) and (5.6) are evaluated. Each of the input variables has a set of values. The states of each variable are shown in Table 5.2.

$$P(Navigation\ Area\ |Traffic,\ Restricted\ Waters) \tag{5.4}$$

$$P(Visibility\ |Daytime,\ Fog,\ Sky,\ Rain) \tag{5.5}$$

$$P(Weather\ |Visibility,\ Sea\ State,\ Wind\ Speed,\ Swell) \tag{5.6}$$

These dynamic variables change during the voyage. The National Meteorological Office of Greece provides data about the current and the future weather conditions. As far as the geomorphological characteristics and the environmental sensitive areas in the Aegean Sea are concerned, the required data are collected from the properly constructed database. The vessel density per navigation area is calculated using the AIS vessels signal. In order to train the dynamic Bayesian network, real weather conditions data and the trajectories of the vessels crossing the Aegean Sea are utilized.

Table 5.2 States of dynamic model

Node	States			
Accident probability	low	moderate	high	
Fog (km)	fog	mist	haze	not fog
Navigation area	low	moderate	high	
Rain (mm)	no rain	light	moderate	high
Restricted waters	yes	no		
Sea state (m)	calm	moderate	rough	high
Sky (% cloud cover)	clear	partly cloudy	cloudy	
Swell (km)	low	moderate	heavy	
Time of day	day	night		
Traffic	low	moderate	high	
Visibility	good	moderate	poor	
Weather conditions	good	moderate	severe	extreme
Wind speed (kn)	calm	breeze	strong	gale

5.2.2 Consequences Estimation

Consequences of the considered marine accidents might affect humans, ships, the environment, industry, and also the waterways. In the context of the present work only the environmental impact has been taken into consideration. Environmental damage, clean-up costs and socioeconomic effects are basically associated with the environment, and mostly with oil spills [19]. The first cost group refers to oil spills that may impact the environment causing physical smothering of organisms, chemical toxicity or ecological changes [15]. As far as the second cost group is concerned, clean-up costs of an eventual spill, research costs and other specific costs, such as the loss of cargo and/or vessels and the repairs needed after accident are considered. According to [3, 10, 19, 29], the main factors influencing the clean-up cost of oil spills are related to (a) the type of oil, (b) the location, (c) the weather and sea conditions, (d) the amount of spilled oil together with the rate of spillage, and (e) the clean-up response. Finally, the socioeconomic effects consist of property damage and income losses [22].

The total cost of an oil spill can be derived by using several different methods. The model proposed by Kontovas et al. [19] is the one that is considered the most appropriate for this work. Therefore, the mathematical framework for estimating the total cost of an oil spill is stated in (5.7).

$$S = 51.432 \times V^{0.728} \tag{5.7}$$

where S represents the consequences measured in U.S. dollars and V is the volume of oil spilled measured in metric tons. The considered cases of V factor are the following three: zero, mean and total outflow. Zero outflow implies no consequences, mean outflow denotes that approximately half of cargo is lost, while in total outflow all the cargo is lost.

5.3 Historical Trajectories Preprocessing and Clustering

This component (preprocessing and clustering) is used to provide the policy recommendation tool with a set of statistics for respective vessel trajectories and uses long-term marine traffic information. At a first stage historical trajectories are read from the HERMES DB [26, 27]. HERMES DB is a database software that can handle trajectories. For each unique vessel in the dataset the respective vessel characteristics (size, ship type, flag, etc.) are retrieved from a vessel database.

At the functional core of this component, the following analysis steps are conducted:

- For each vessel, the Bayesian Network that estimates the vessel risk is employed to infer a single accident probability per vessel in the dataset.
- A set of trajectory-specific features is extracted, mostly based on velocity, initial and final position, destination, speed deviation, vessel type and flag. This is fed as input to a hierarchical clustering method in order to extract groups of similar (in terms of the adopted features) trajectories.

The second step (clustering) comprises the task of grouping objects based on their spatial and temporal similarity [18]. Therefore, the goal of this procedure is to group similar vessel trajectories into homogeneous categories, in terms of similar trajectory attributes. The output of this procedure can be considered as a set of channels of similar trajectories (Fig. 5.5). Each set of channels is mapped to a single trajectory cluster (Dominant Trajectory Cluster) that can be considered as an aggregated representative of similar trajectories.

The usefulness of this functionality is twofold:

- Provision of a better visualization of large trajectory datasets, since each group of similar trajectories can be visualized by its respective "centroid"
- Extraction of a more efficient and compressed representation of trajectory clusters: instead of storing and analyzing attributes on the whole set of initial trajectories, the computations are performed over the aggregates of each cluster. Towards this end, for each cluster an aggregate attribute representation is stored (e.g. the histogram of ship sizes) instead of the initial (raw) characteristics.

In order to achieve the clustering of the trajectories, each trajectory is represented by a set of attributes: average speed, standard deviation of speed, minimum and maximum longitude and latitude, moments computed over the vessel direction angle and distance metrics. Next, a fast hierarchical clustering implementation is incorporated

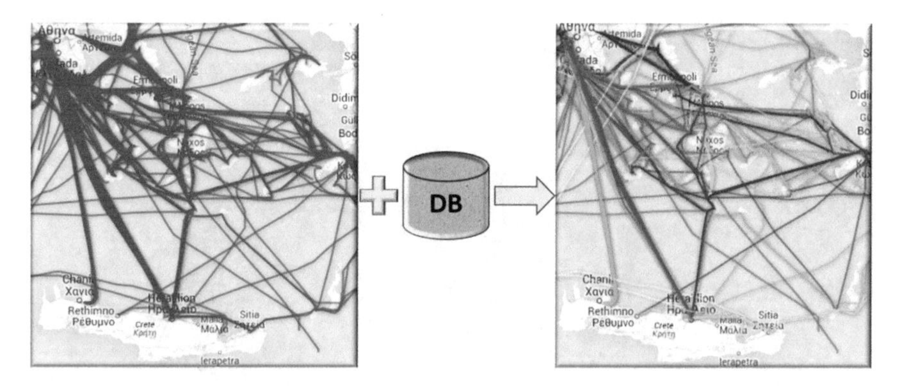

Fig. 5.5 Trajectory clustering

on the adopted feature space [25]. The resulting enriched dataset is stored in an intermediate Fusion Table which is then used as input by the policy recommendation tool to execute spatial queries and extract aggregated statistics.

5.4 Decision Support Tool

In this section the decision support tool is presented in detail. More specifically, the functionalities of the on-line tool, the technologies used, and data analytics are described.

5.4.1 Description of Tool Functionalities

The decision support tool is implemented as a user-friendly web application with a major functionality the visualisation of the historical trajectories and their attributes (e.g. cluster labels, dynamic and static risks extracted by the aforementioned modules). This application accepts spatial input from the user, executes a spatiotemporal query to the system database and finally displays the results with several descriptive diagrams.

In Fig. 5.6 the web application is shown. The web page consists of a map area, a toggle button area, a selection area, and a results area (tabs) where statistical diagrams are displayed. First, in the “Toggle Layers” section the user can select one or more environmental layers (as decribed in Subsect. 5.5.1). Also, the user can select the layer of Dominant Trajectories Clusters (as described in Sect. 5.3). The selected layers are shown on the top of the map. In the selection area, the user can:

- Save results (for further analysis)
- Select saved results (for display)

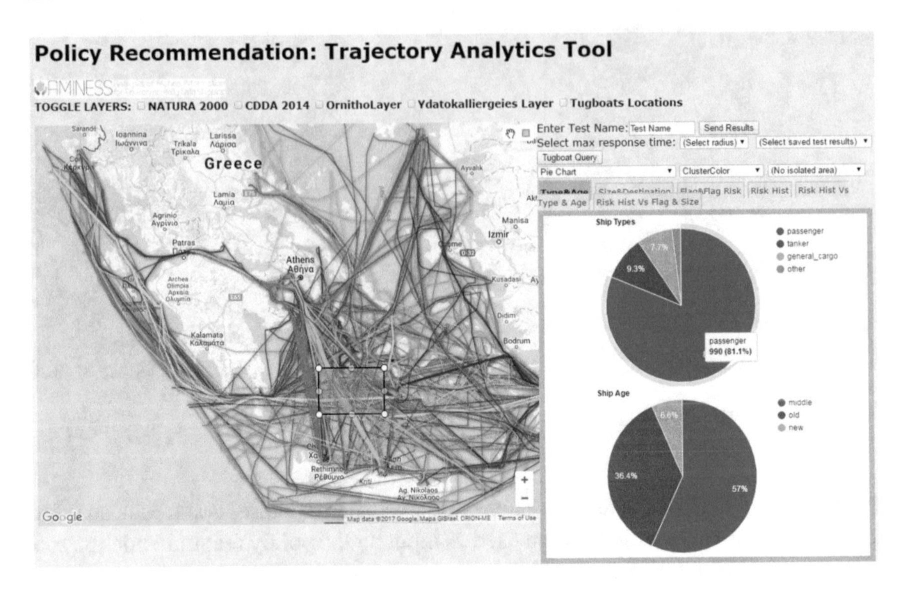

Fig. 5.6 Policy recommendation tool

- Select the type of chart (pie, timeline chart 1 or 2) to display in the current tab (described in Subsect. 5.4.3)
- Select the color scheme of vessel trajectories. Color by dominant cluster (default), risk, risk of accident (described in Sect. 5.2)
- Select dataset of vessel trajectories: normal (default), with restricted areas as described in Subsect. 5.5.3

In the map area two tools are available: pan tool (to move the map) and rectangular tool. Selecting the rectangular tool the user can draw a rectangular area over the map. This selected area defines the spatial filter.

The results area includes six tabs. Each tab contains (aggregated) statistical diagrams of the results. The diagrams of the tabs are described in Subsect. 5.4.3.

The main focus of the visualization procedure is the extraction of aggregated statistics for a given spatial filter (bounding rectangle). These statistics are displayed through respective diagrams. In general, the spatiotemporal variability of the attributes of the selected trajectories is portrayed in those diagrams. The trajectories are selected through a spatiotemporal query on the system database defined by the bounding box of the user input.

In our system the datasets are stored in Google Fusion Tables (see Subsect. 5.4.2). A snapshot of our dataset is shown in Fig. 5.7. For every vessel trajectory the various attributes are shown (e.g. ship type, age, flag, risk etc.) together with the geometry of it. The trajectory coordinates are stored as KML LineString (see Subsect. 5.4.2). This storage schema makes the trajectory retrieving process via intersection queries rather simple and fast. In Fig. 5.8 a sample record (Fusion Table row) view is displayed. Beside the row attributes the vessel trajectory is displayed on a map.

bCompressed

Imported at Thu Sep 25 08:54:28 PDT 2014 from bCompressed.csv

Add Attribution - Edited on 2014 November 17

File Edit Tools Help | Rows 1 | Cards 1 | Map ClusterColor | Map RiskColor | Map RiskAccidentColor

Filter No filters applied

1-100 of 9456

ID	Port	Cluster	ClusterColor	Type	Date	Vel	ShipAge	ShipFlag	ShipFlagRisk	ShipSize	Risk	RiskAccident	RiskColor	RiskAccidentColor	Path
239882000 - 18	PERAMA	CL0	#00ad60	other	2009-06-15 11:30:22	135.464427185	middle	Greece	low	medium	0.24	0.128	#0074ff	#0000ff	KML...
239882000 - 19	PIRAIEVS	CL1	#020820	other	2009-06-16 03:17:07	65.6110482525	middle	Greece	low	medium	0.24	0.128	#0074ff	#0000ff	KML...
239692000 - 418	PIRAIEVS	CL2	#0362e0	passeng	2009-08-23 13:10:36	428.869442816	middle	Greece	low	small	0.26	0.316	#0088ff	#00c0ff	KML...
239692000 - 419	PIRAIEVS	CL3	#04bda0	passeng	2009-08-23 16:40:02	495.730282688	middle	Greece	low	small	0.26	0.316	#0088ff	#00c0ff	KML...
240694000 - 119	PERAMA	CL1	#020820	other	2009-07-10 20:37:07	12.5430826026	middle	Greece	low	small	0.24	0.023	#0074ff	#000096	KML...
240694000 - 118	PIRAIEVS	CL1	#020820	other	2009-07-10 00:03:36	5.90072244682	middle	Greece	low	small	0.24	0.023	#0074ff	#000096	KML...
240694000 - 115	PERAMA	CL3	#04bda0	other	2009-07-08 09:40:30	7.98592056043	middle	Greece	low	small	0.24	0.023	#0074ff	#000096	KML...
240694000 - 117	PERAMA	CL0	#00ad60	other	2009-07-09 20:40:57	8.22155564975	middle	Greece	low	small	0.24	0.023	#0074ff	#000096	KML...
240694000 - 113	PERAMA	CL0	#00ad60	other	2009-07-07 21:45:36	9.63706444989	middle	Greece	low	small	0.24	0.023	#0074ff	#000096	KML...
240694000 - 112	PIRAIEVS	CL1	#020820	other	2009-07-07 05:00:57	6.83910167886	middle	Greece	low	small	0.24	0.023	#0074ff	#000096	KML...
237470200 - 9	AYIOS NIKOLAOS	CL2	#0362e0	tanker	2009-06-05 03:57:26	13.1592342575	old	Greece	low	small	0.53	0.195	#93ff63	#0044ff	KML...
237470200 - 8	PERAMA	CL4	#061860	tanker	2009-06-04 09:21:35	16.3736102536	old	Greece	low	small	0.53	0.195	#93ff63	#0044ff	KML...
237470200 - 5	ELEVSIS	CL3	#04bda0	tanker	2009-06-03 02:52:52	14.0908056065	old	Greece	low	small	0.53	0.195	#93ff63	#0044ff	KML...
237470200 - 4	PERAMA	CL4	#061860	tanker	2009-06-02 04:00:14	9.81341726358	old	Greece	low	small	0.53	0.195	#93ff63	#0044ff	KML...
237470200 - 7	PIRAIEVS	CL1	#020820	tanker	2009-06-04 04:56:44	5.18657122251	old	Greece	low	small	0.53	0.195	#93ff63	#0044ff	KML...
237470200 - 6	PERAMA	CL1	#020820	tanker	2009-06-03 16:00:03	10.0912446584	old	Greece	low	small	0.53	0.195	#93ff63	#0044ff	KML...

Fig. 5.7 Fusion Table (sample) of vessel trajectories and attributes

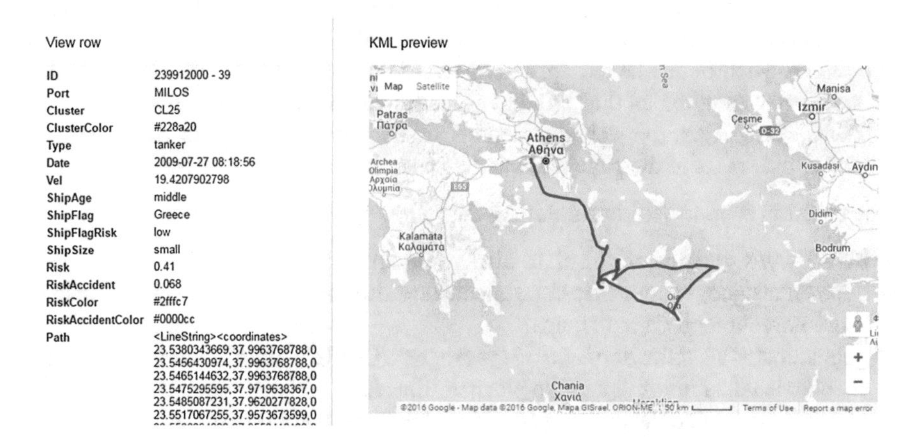

Fig. 5.8 Fusion Table row view

In the web tool, after the user has selected the dataset and the rectangle area, an intersection is performed (on the server side) on the trajectories of the Fusion Table. The trajectories that are contained or intersected by that rectangular are selected and returned to the interface together with their attributes. After the retrieval of the intersected trajectories, the respective attributes are aggregated and presented as statistical diagrams to the user. These diagrams are discussed in Subsect. 5.4.3.

Also, the results of each query can be saved. First, the user enters the test name in the text area *"Enter Test name"* and after pressing the button *"Send Results"* the data are stored in a Fusion Table. The data stored include the test date, the Fusion table (of the vessel trajectories) used, the spatial extent of the query, the intersected trajectories and its attributes. These data can be further processed or retrieved and displayed in the web tool using the drop-down menu *"Select saved test results"*. Next, the technologies used in this web tool will be presented followed by the data analytics.

5.4.2 Technologies Used

For data visualization in the context of a web application environment, the following technologies are adopted:

Fusion Tables: Google Fusion Tables (Google [8]) is a web application provided by Google for data management. Fusion tables can be used for gathering, visualizing and sharing data tables. Data are stored in multiple tables that Internet users can view, edit and download. The web application provides means (APIs) for visualizing data using pie charts, bar charts, lineplots, scatterplots, timelines, and geographical maps. Example of a Fusion Table is shown in Fig. 5.7, while Fig. 5.8 shows a Fusion Table row view.

KML: Keyhole Markup Language (KML, Google [9]) is a file format used to display geographic data in Internet-based maps or Earth browsers and is based on the XML standard. KML elements are stored in Fusion Tables as record attributes. In our datasets KML *LineString* represents a vessel trajectory.

Ajax: Asynchronous JavaScript and XML (Ajax, Wikipedia [30]) is a set of client-side web technologies that provide asynchronous functionality to web applications. Using Ajax, the web applications are able to exchange data with a server, and update parts of the page without the need for a complete web page reloading.

The decision support tool is employing the following (Javascript) libraries:

- https://www.google.com/jsapi: to display Google maps.
- http://maps.google.com/maps/api/js?sensor=false&libraries=drawing: to enable user interaction on Google maps.
- google.load("visualization", "1", {'packages': ['table', 'map', 'corechart']}): to enable visualization API and the piechart, linechart, and histogram packages.
- http://geoxml3.googlecode.com/svn/branches/polys/geoxml3.js: this library renders KML on the Google Maps JavaScript API.

5.4.3 Data Analytics

The data analytics (statistics), in the results area of the web tool, are organized into six tabs. Each tab contains two separate parts (different diagrams) that display the variability of attributes of the selected trajectories (Fig. 5.6). More specifically the attributes analyzed by this software are the following:

- Ship type
- Ship age
- Ship size
- Port of destination
- Ship flag
- Ship flag risk (risk due to the ship flag)

- Histogram of risk (histogram of aggregated estimated risks)
- Risk distribution due to ship attributes: this function can be used as an explanatory analysis of the distribution of risk with respect to certain attributes of the ships, namely:
  - Histogram of risk with respect to the ship type
  - Histogram of risk with respect to ship age
  - Histogram of risk with respect to ship flag
  - Histogram of risk with respect to ship size

For the visualization of the statistical results the following types of charts are employed:

Pie chart: Depicts the percentage (%) ratio of vessels per attribute category (Fig. 5.9)

Timeline chart 1: The vertical axis shows the absolute number of ships (Fig. 5.10)

Timeline chart 2: The vertical axis shows the percentage of vessels (normalized for each day) (Fig. 5.11)

Histogram bar and timeline chart: A bar graph where the horizontal axis shows the number of ships and the vertical axis the risk (frequency of discrete risk values). Timeline chart displays the variability of maximum and average risks in time (Fig. 5.12)

Stacked histogram: Each bar is divided into sections according to the sizes of the attribute classes. This diagram is used to display the risk distribution of the specific vessel attribute with respect to the attribute classes (Fig. 5.13)

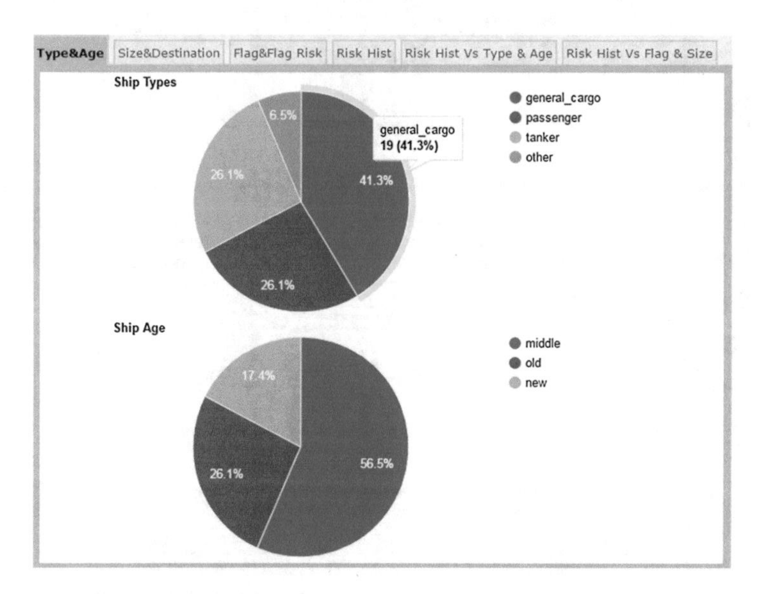

Fig. 5.9 Example of Pie Charts (ShipType, ShipAge)

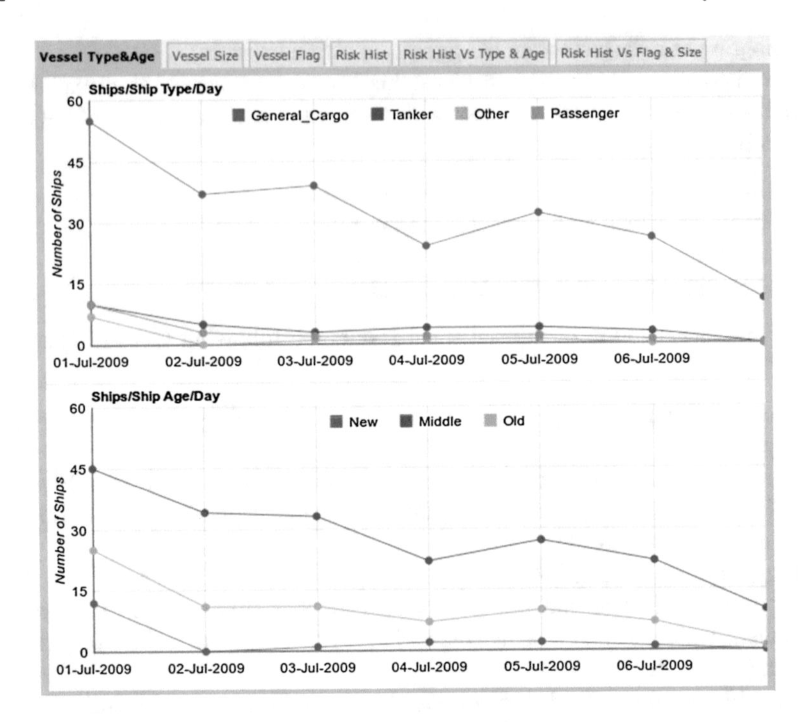

Fig. 5.10 Example of vessel type and age timeline (absolute)

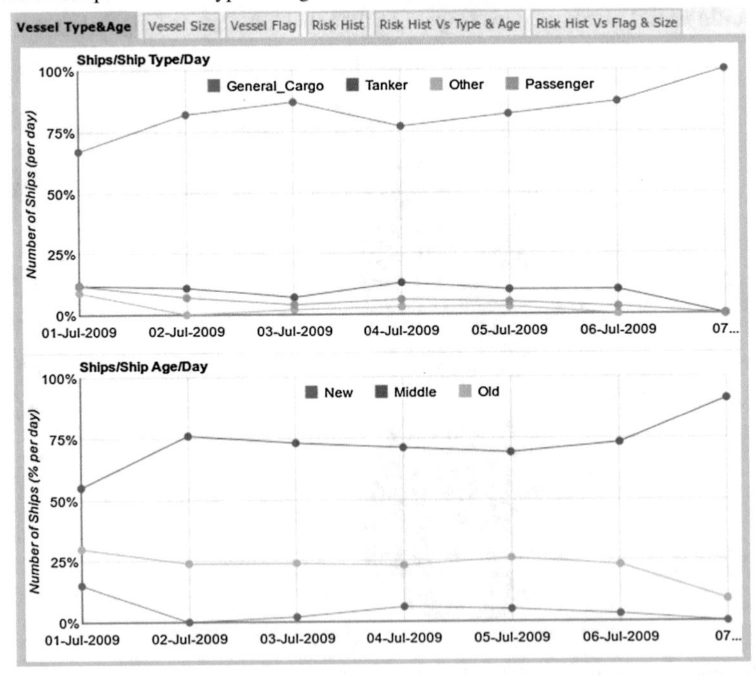

Fig. 5.11 Example of vessel type and age timeline (%)

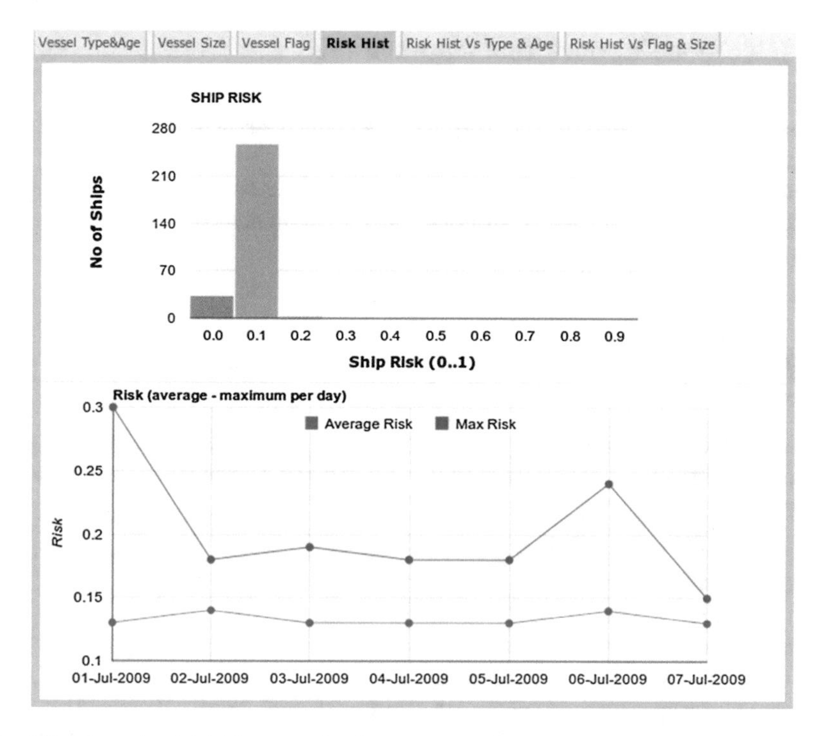

Fig. 5.12 Example of risk histogram timeline (avg, max)

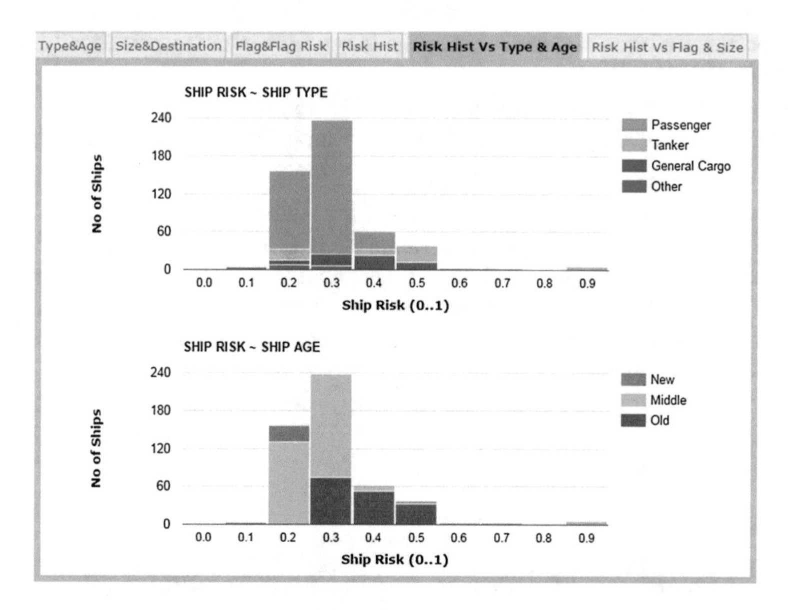

Fig. 5.13 Example of risk histogram per ShipType and per ShipAge

5.5 Decision Support Tool Applications

In this section the capabilities of the policy recommendation tool are demonstrated in applications where this tool has been employed. First, queries in environmentally sensitive areas are presented. Next, an example of marine traffic monitoring for environmental safety is described followed by a scenario of restricted areas in marine traffic.

5.5.1 Queries in Environmentally Sensitive Areas

The capabilities of the policy recommendation tool are extended by including layers of environmentally sensitive areas of the Aegean and Ionian Seas. The following environmentally sensitive layers are included:

Natura 2000: this is an ecological network of protected areas set up to ensure the survival of Europe's most valuable species and habitats, and includes both terrestrial and marine sites (European Environment Agency [5]),

CDDA 2014: the Common Database on Designated Areas (CDDA) is a data bank for officially designated protected areas in Europe such as nature reserves, protected landscapes, national parks etc. (European Environment Agency [4]),

Ornithological: this layer contains the most important bird nesting areas of Greece (as provided by the Hellenic Ornithological Society). These are sites particularly important for bird conservation and selected on the basis of internationally agreed standard criteria, and

Ydatokalliergeies: this layer includes the fishery areas of the Aegean and Ionian Seas (as provided by the Greek Ministry of Rural Development and Food)

At the web tool the user can select (at the toggle layers area) to view any of these layers. An example of a query in an area of the Natura 2000 layer is presented in Fig. 5.14. In this case, clicking on any Natura 2000 area launches an information box with the *ID* and *Description* attributes of that area. Also, a *Query* button is included in that info box. Clicking on that button automatically executes the query. The geometrical limits of the spatial query are defined by the minimum bounding box of the selected environmental area. As soon as the server responds, the intersected trajectories and the aggregated statistics for this specific Natura 2000 area are displayed in the web tool. Finally, the user can select a test name and send the results to a Fusion table for storage and further processing.

5.5.2 Marine Traffic Monitoring for Environmental Safety

In this application (Giannakopoulos et al. [6]) aggregated statistics are extracted from seven areas of major interest in the Greek marine territory employing the decision

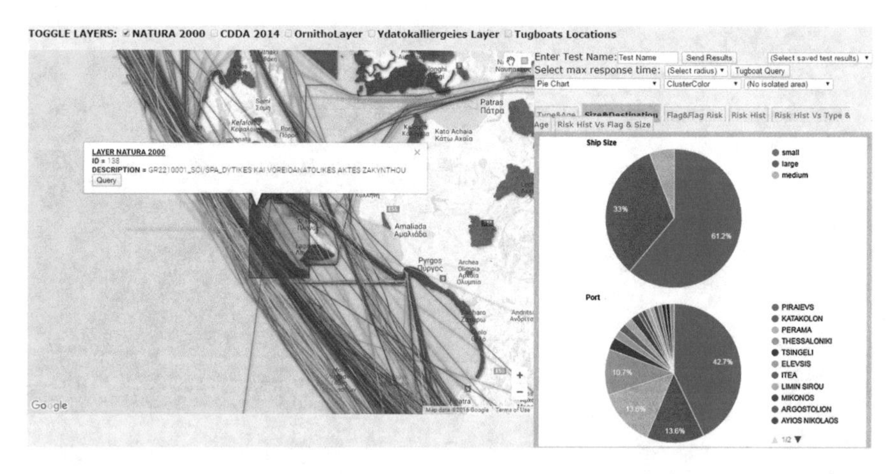

Fig. 5.14 Query in a natura 2000 Area

Table 5.3 Monitoring areas and general sectors

Area Name	Sector	Sector ID
Kasos	Southern Aegean-Cretan Sea	1
Kythera	Southern Aegean-Cretan Sea	1
Zakynthos	Ionian-Patras Gulf	2
Amorgos	Central Aegean-Corinth Gulf	3
Koufonissia	Central Aegean-Corinth Gulf	3
Lemnos	Northern Aegean	4
Sporades	Northern Aegean	4

support tool. The objective of this research is to study the correlation between spatial environmental importance and marine traffic risk.

These areas in the Greek marine territory have been selected due to their environmental importance. Criteria for this selection are: area must either be protected (e.g. NATURA 2000) or shelters highly important species. In this research areas have been selected from the NATURA 2000 layer. In Table 5.3 the names of these areas are displayed together with their respective general sectors and sector IDs.

For each area and its corresponding sector the decision support tool has been employed to detect (using spatial query) the contained trajectories and display their aggregated statistics, focusing on the average vessel risk. The spatial extent of the general sectors are presented in Fig. 5.15. For the selected areas the spatial extent is defined by the minimum bounding box of the corresponding NATURA 2000 area. Queries are performed in the general sectors and specific areas (as described in Subsect. 5.5.1). Test results are stored in Fusion Tables and exported. Next, the exported data are used for computing the following statistical measures.

Risk (A) : Area risk = average risk for all vessels in that area

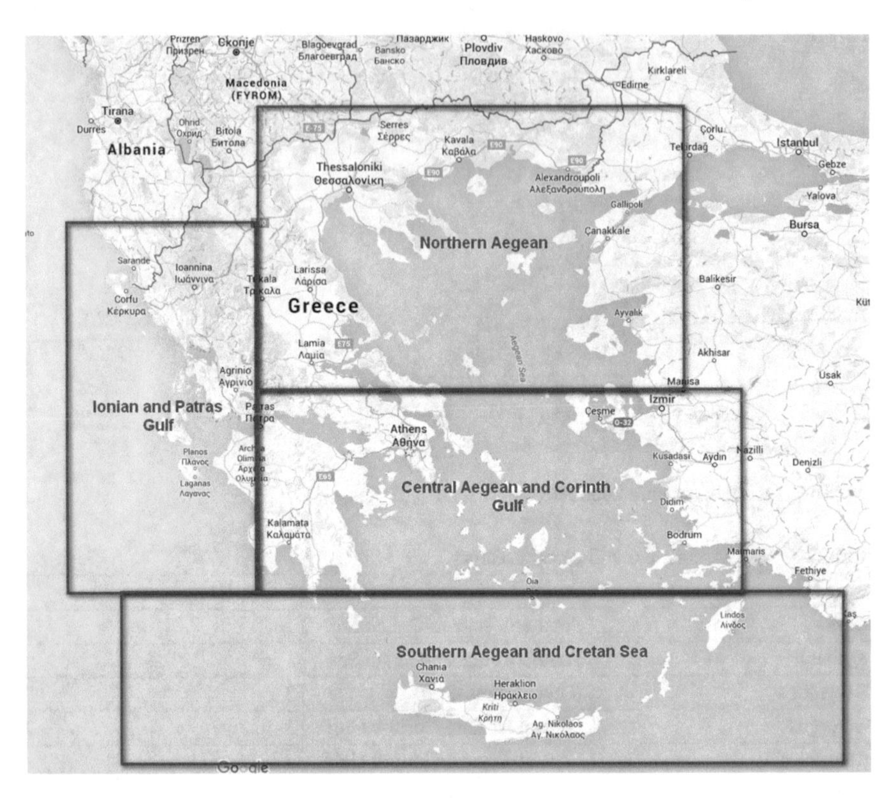

Fig. 5.15 Spatial extent of general sectors

Risk (S) : Sector risk = average risk for all vessels in that sector
Risk Overall : average risk for all vessels in the database
%R>x : number of vessels (in area *A* or sector *S*) with risk greater than *x*

The aggregated risk measures are presented in Table 5.4. In this table the following comparisons are depicted:

- Comparison between each sector risk, each included area risk and the overall risk
- Comparison between the percentages of vessel trajectories, with risk greater than specific thresholds (0.4, 0.5 and 0.6), in each sector and each included area

The results of Table 5.4 demonstrate that the average risk of all areas considered, is greater than the corresponding general sector's risk. Additionally, the percentage of trajectories belonging to the three high-risk-zones is almost always extensively greater than the respective percentages in the sector's aggregates. The above results indicate that these environmental sensitive areas have a great probability to experience a marine accident due to vessel traffic with high risk.

This application of marine traffic monitoring for environmental safety can be expanded by studying the temporal variability of marine traffic through environmental sensitive areas and useful conclusions can be derived for policy makers. For

Table 5.4 Aggregated risk results of areas of interest and general sectors

Area	Sector	Risk			%$R > 0.4$		%$R > 0.5$		%$R > 0.6$	
(A)	(S)	A	S	Overall	A	S	A	S	A	S
Kasos	1	0.38	0.35		24	20	10	7	4	2
Kythera	1	0.37	0.35		22	20	10	7	3	2
Zakynthos	2	0.36	0.35		23	22	10	11	2	1
Amorgos	3	0.38	0.34	0.34	40	30	24	16	2	2
Koufonissia	3	0.36	0.34		36	30	20	16	1	2
Lemnos	4	0.40	0.36		32	24	13	12	5	2
Sporades	4	0.40	0.36		43	24	20	12	3	2

example, a case study could examine (for a Natura 2000 area) the temporal variability of risk for ship types *general cargo* and *tanker* and ship ages *old* and *middle*.

5.5.3 *Influence of Restricted Areas in Marine Traffic*

In this section the influence of restricted areas in marine traffic is examined. Two case studies are considered where marine traffic limitation is imposed in specific areas. An illustration of using the monitoring tool for implementation of if-then-scenarios in maritime traffic is described in Gyftakis et al. [11].

The (common) scenario in each case study includes complete vessel traffic restriction in an area specified by a rectangular box (spatial filter) and for the time length of the complete original dataset (no temporal filter). Using the spatial filter a query is performed in the complete dataset and the interesected trajectories (with their attributes) are returned. These trajectories represent the vessels that need to be redirected due to traffic restriction in that area. For this purpose, a graph-based optimal route extraction algorithm is utilized in a batch mode on the restricted trajectories (Makrygiorgos et al. [23]). The optimal route algorithm is based on a search grid with interconnected nodes. The nodes of that search grid that are contained in the restricted area are deactivated. The optimal route selection is performed on the modified search grid. This modification guarantees that the new trajectories will divert from the restricted area. The dataset of the produced trajectories together with the rest trajectories of the original dataset are stored in a Fusion Table. The user can select this alternative dataset in the selection area of the web tool interface (e.g. *Cythera Isolated Area*).

For comparison reasons, the scenario contains the examination of marine traffic in an area close to the restricted one. Queries are performed in that area (examination area) using the original (without any area restriction) dataset of trajectories and the alternative one (with the area restriction).

Policy makers can obtain valuable information from the comparison of the results of those queries. For example, if a major marine accident happens in a specific area and that area must be restricted, what are the consequences (of marine traffic) at nearby areas? In another example, if traffic of specific type of vessels must be restricted in an area, what are the optimal time periods? In each case, the proposed decision support tool delivers analytic data and statistics to support a policy with objective arguments.

In the following subsections two case studies with restricted areas in the Aegean Sea are described; one at Cythera Strait and another one at Kafireas Strait.

5.5.3.1 Restricted Area: Cythera Strait

In this example the restricted area is the Cythera strait that is formed by the southeastern peninsula of the Peloponnese and the islands of Elafonissos and Cythera. This area represents one of the most dangerous navigational areas in the Mediterranean.

The event of without restriction is displayed in Fig. 5.16. In more detail, the examination area is a rectangular area nearby the Cythera strait. Marine traffic is dense through the Cythera Strait and through the examination area. The original dataset has been used for the spatial query.

The results of the query are displayed in Fig. 5.16. In this figure two characteristic diagrams are shown, the risk histogram versus ship type and ship age. In these diagrams the largest frequency is at risk 0.3 with 14 ships (4 passenger and 10 cargo ships). The maximum risk (0.9) is due to 2 middle-aged tanker ships.

Next, the event of area restriction is examined. In Fig. 5.17 the restricted area is displayed in red (Cythera strait). As described in Subsect. 5.5.3 imposing this area restriction a modified dataset is produced. The dataset used in this case is the *Cythera Isolated Area*. Using the same examination area as before a query is performed and the results are displayed in Fig. 5.17. The diagrams displayed are the risk histogram versus ship type and ship age.

Overall the risk histograms follow the same distribution as in the event of without restriction. In the diagrams of the restriction event, the largest frequency is at risk 0.3 with 24 ships (13 passenger and 11 cargo ships) compared to 14 (4 passenger and 10 cargo ships) before. Hence, the increase in the number of ships (for risk 0.3) is mainly due to passenger ships diversion. Also, the number of ships for maximum risk (0.9) remains the same as before.

The conclusions are that cargo ships and tankers avoid the Cythera strait and the restriction increases traffic of passenger ships in nearby areas.

5.5.3.2 Restricted Area: Kafireas Strait

In the second example the restricted area is the Kafireas strait between the island of Euboea and the island of Andros (in earlier times this strait was one of the most dangerous navigational areas in the entire Aegean).

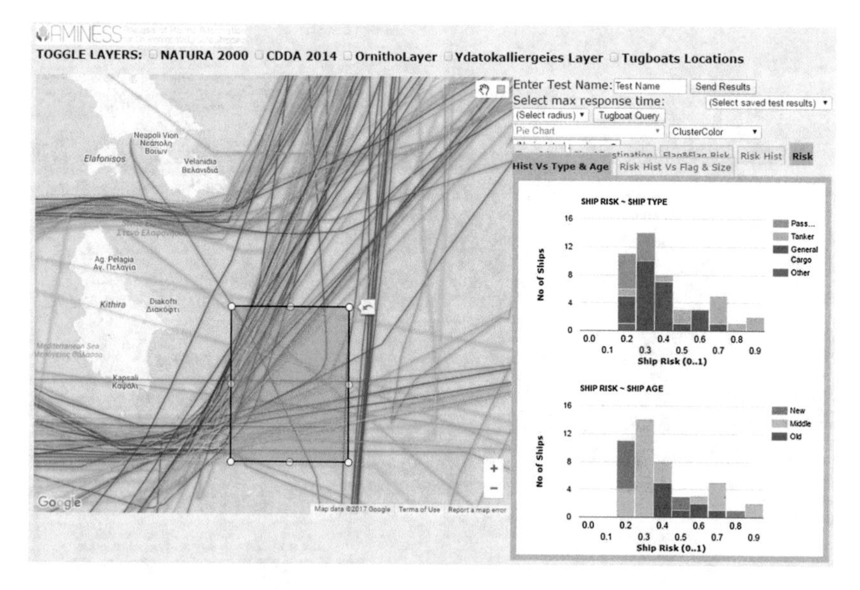

Fig. 5.16 Area near Cythera Strait without restriction

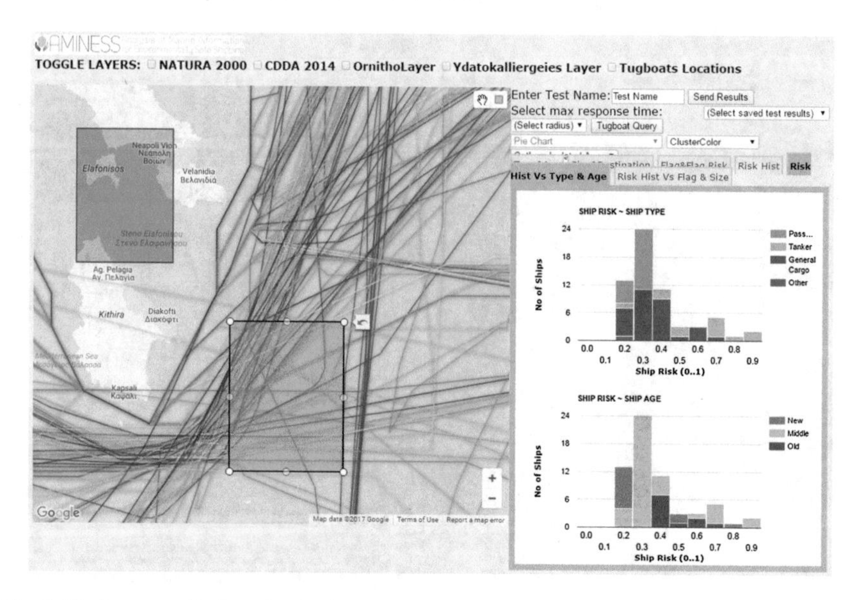

Fig. 5.17 Area near Cythera Strait with restriction

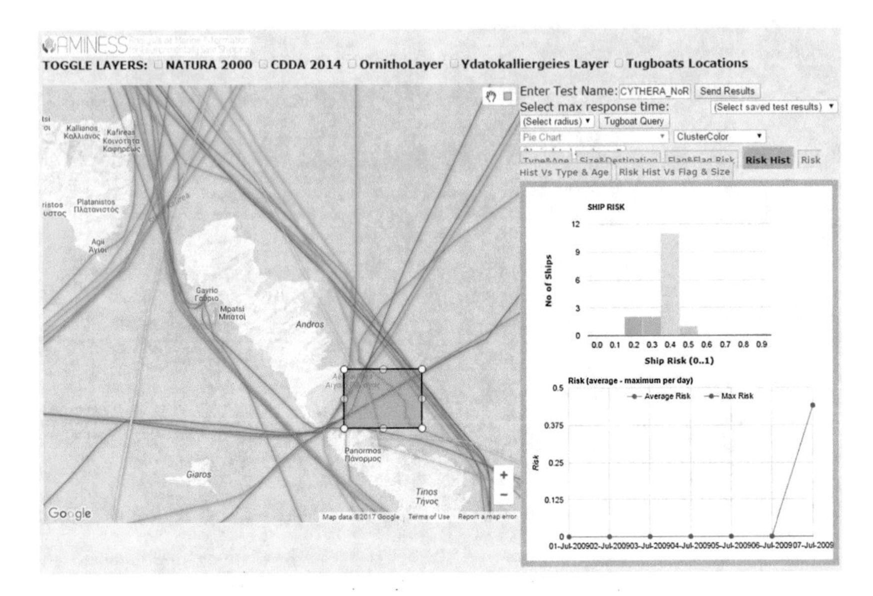

Fig. 5.18 Area near Kafireas Strait without restriction

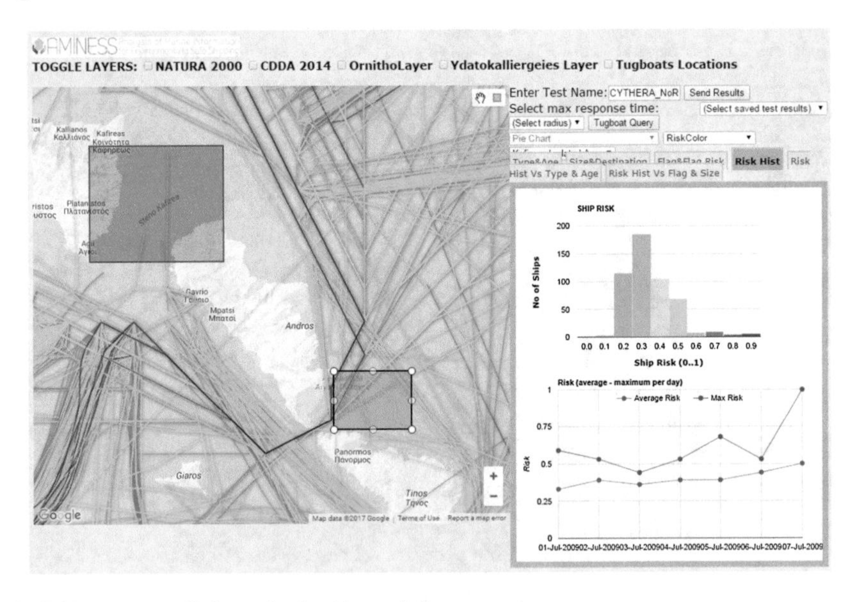

Fig. 5.19 Area near Kafireas Strait with restriction

The restricted area is shown (red area) in Fig. 5.19 while in Fig. 5.18 an examination area is selected close to the restricted one and away of the major ship routes. In the event of without restriction (Fig. 5.18) the original dataset has been used for the spatial query.

The results of the query are displayed in Fig. 5.18. The diagrams displayed in this figure are the risk histogram and the risk timeline (daily average, daily maximum). In the risk histogram the largest frequency happens at risk 0.4 with numbers of ships 11. There are no ships for frequencies greater than 0.5. In the timeline diagram there are no ships except the last day.

Next, the event of restriction in Kafireas strait is explored. Following the procedure in Subsect. 5.5.3 a modified dataset is produced (*Kafireas Isolated Area*). Using this dataset and the same examination area as before a query is performed and the results are displayed in Fig. 5.19. The diagrams displayed in this figure are the risk histogram and the risk timeline (daily average, daily maximum).

In the histogram diagram the largest frequency happens at risk 0.4 with number of ships 185. For risk frequencies greater than 0.5 there are totally 26 ships. The timeline diagram contains ships for every day in contrast to the timeline of the no restricton event. The maximum risk line is almost every day higher than the 0.5 value.

The conclusions are that imposing restriction in Kafireas strait can generate high marine traffic in areas with minimal marine traffic. This incease produces high risk traffic every day of the week.

5.6 Conclusions—Future Plans

In this chapter a decision support tool employing Bayesian risk framework for environmentally safe shipping is presented. The objective of the presented tool is to handle information from vessel trajectories in order to support policy actions for environmental risk reduction in Greek Seas. The proposed system includes preprocessing, clustering of trajectories (based on their spatial and temporal similarity) and risk assessment using probabilistic models. The latter was accomplished through the construction of a Bayesian network able to predict the probability of marine accident (i.e. collision, contact and grounding) using dynamic information about the navigation area and weather conditions. The functionalities, technologies used and data analytics of the tool are discussed. Applications of the decision tool are presented in areas such queries in protected areas, marine traffic monitoring for environmental safety and influence of restricted areas in marine traffic. Analysis of the results, especially risk diagrams, demonstrates the significance of using the developed tool for maritime policy makers.

The presented functionalities of the decision support tool deliver important information for the expansion of this system. For a lot of types of what-if-scenarios modules can easily assembled and tested. For example, the study of maritime traffic restrictions can produce important information on usefulness and side effects of such restrictions. Temporal queries can be explored and possible links with accident data as well.

A step forward in risk assessment is to integrate both static and dynamic risk models with the consequences model, in order to consider the exact extent of the consequences.

The trajectories classification methodology can be expanded to identify spatiotemporal trends and abrupt changes in the maritime trajectories. This classification scheme can be used as an effective data modeling technique. Each cluster will contain cumulative information (in the form of histograms of statistical data) characterizing the respective trajectories. This information will be stored in the Fusion table used by the proposed tool, that will be able to retrieve aggregated statistics by cluster paths, instead of multiple statistics for all similar routes. Hence, this technique can lead to an efficient data compression resulting in faster retrieving time for analysis and displaying in real time.

Acknowledgements This work was carried out in the framework of the project "AMINESS: Analysis of Marine Information for Environmentally Safe Shipping" that was co-financed by the European Fund for Regional Development and by Greek National funds through the operational programs "Competitiveness and Entrepreneurship" and "Regions in Transition" of the National Strategic Reference Framework—Action: "COOPERATION 2011 Partnerships of Production and Research Institutions in Focused Research and Technology Sectors".

References

1. Bingham, P., Koch, L.: Liner shipping in the european union. Technical report, The World Shipping Council and IHS Global Insight (2009)
2. Det Norske Veritas (DNV): Formal safety assessment—Large passenger ships, annex ii: Risk assessment large passenger ships—Navigation. Technical report, Det Norske, Veritas (2003)
3. Etkin, D.: Estimating clean-up costs for oil spills. In: International Oil Spill Conference, American Petroleum Institute. Washington, DC (1999)
4. European Environment Agency: Nationally designated areas (CDDA) (2016a). http://www.eea.europa.eu/data-and-maps/data/nationally-designated-areas-national-cdda-10
5. European Environment Agency: Natura 2000 data—The European network of protected sites (2016b). http://www.eea.europa.eu/data-and-maps/data/natura-7
6. Giannakopoulos, T., Gyftakis, S., Charou, E., Perantonis, S., Nivolianitou, Z., Koromila, I., Makrygiorgos, A.: Long-term marine traffic monitoring for environmental safety in the Aegean Sea. In: 36th International Symposium on Remote Sensing of Environment. Berlin, Germany, May 2015
7. Giannakopoulos, T., Vetsikas, I., Koromila, I., Karkaletsis, V., Perantonis, S.: Aminess: a platform for environmentally safe shipping. In: 7th International Conference on PErvasive Technologies Related to Assistive Environments. Rhodes, Greece, May 2014
8. Google: About Fusion Tables (2016a). https://support.google.com/fusiontables/answer/2571232?hl=en
9. Google: Keyhole Markup Language (2016b). https://developers.google.com/kml/documentation/kml_tut
10. Grey, C.: The cost of oil spills from tankers: an analysis of iopc fund incidents. In: International Oil Spill Conference. vol. 1, pp. 41–47. American Petroleum Institute (1999)
11. Gyftakis, S., Giannakopoulos, T., Makrygiorgos, A., Charou, E., Perantonis, S., Koromila, I., Nivolianitou, Z.: A maritime data analytics platform for policy recommendation. In: 6th International Conference on Information, Intelligence, Systems and Applications. Corfu, Greece, July 2015
12. Hanninen, M.: Bayesian networks for maritime traffic accident prevention: Benefits and challenges. Accid. Anal. Prev. **73**, 305–312 (2014)

13. Hanninen, M., Kujala, P.: Bayesian network modeling of port state control inspection findings and ship accident involvement. Expert Syst. Appl. **41**(4), 1632–1646 (2014)
14. IMO: International maritime organization. msc 83/inf.2. formal safety assessment: Consolidated text of the guidelines for formal safety assessment (fsa) for use in the imo rule-making process (2009). (msc/circ.1023mepc/circ.392)
15. ITOPF: Tip 13: Effects of oil pollution on the marine environment. Technical report, International Tanker Owners Pollution Federation (2014)
16. Jensen, J., Soares, C., Papanikolaou, A.: Methods and tools. In: Papanikolaou, A. (ed.) Risk-Based Ship Design: Methods, Tools and Applications, pp. 213–231. Springer, Berlin, Heidelberg (2009)
17. Jiacai, P., Qingshan, J., Zheping, S., Jinxing, H.: An ais data visualization model for assessing maritime traffic situation and its applications. Proc. Eng. **29**, 365–369 (2012)
18. Kisilevich, S., Mansmann, F., Nanni, M., Rinzivillo, S.: Spatio-temporal clustering: a survey. In: Maimon, O., Rokach, L. (eds.) Data Mining and Knowledge Discovery Handbook, 2nd ed, pp. 855–874. Springer Science+Business Media (2010)
19. Kontovas, C., Psaraftis, H., Ventikos, N.: An empirical analysis of iopcf oil spill cost data. Mar. Pollut. Bull. **60**, 1455–1466 (2010)
20. Koromila, I., Nivolianitou, Z., Giannakopoulos, T.: Bayesian network to predict environmental risk of a possible ship accident. In: 7th International Conference on PErvasive Technologies Related to Assistive Environments. Rhodes, Greece, May 2014
21. Koromila, I., Nivolianitou, Z., Giannakopoulos, T., Perantonis, S., Charou, E., Gyftakis, S.: A dynamic model for environmentally safe shipping through the Aegean Sea. In: 6th International Conference on Information, Intelligence, Systems and Applications. Corfu, Greece, July 2015
22. Liu, X., Wirtz, K.: Total oil spill costs and compensations. Marit. Policy and Manage. **33**(1), 49–60 (2006)
23. Makrygiorgos, A., Giannakopoulos, T., Perantonis, S.: Accelerating multi-objective ship routing using a novel grid structure and a simple heuristic. In: 1st International Workshop on Modelling, Computing and Data Handling for Marine Transportation, IISA 2015. Corfu, Greece, July 2015
24. Montewka, J., Ehlers, S., Goerlandt, F., Hinz, T., Tabri, K., Kujala, P.: A framework for risk assessment for maritime transportation systems—A case study for open sea collisions involving ropax vessels. Reliab. Eng. Syst. Saf. **124**, 142–157 (2014)
25. Müllner, D.: fastcluster: Fast Hierarchical, Agglomerative Clustering Routines for R and Python. J. Statist. Softw. **53**(1), 1–18 (2013)
26. Pelekis, N., Frentzos, E., Giatrakos, N., Theodoridis, Y.: Hermes: A trajectory db engine for mobility-centric applications. In: ACM SIGMOD International Conference on Management of Data. Vancouver, Canada, June 2008
27. Pelekis, N., Frentzos, E., Giatrakos, N., Theodoridis, Y.: Hermes: A trajectory db engine for mobility-centric applications. Int. J. Knowl. Based Organ. **5**(2), 19–41 (2015)
28. Silveira, P., Teixeira, A., Soares, C.: Use of ais data to characterize marine traffic patterns and ship collision risk off the coast of portugal. J. Navig. **66**(6), 879–898 (2013)
29. White, I., Molloy, F.: Factors that determine the cost of oil spills. In: International Oil Spill Conference. Vancouver, Canada (2003)
30. Wikipedia: Ajax (2016). https://en.wikipedia.org/wiki/Ajax_(programming)
31. Willems, N., Wetering, H.V.D., Wijk, J.V.: Visualization of vessel movements. Comput. Graph. Forum **28**(3), 959–966 (2009)

Chapter 6
A Decision Support System for the Assessment of Seaports' Security Under Fuzzy Environment

Andrew John, Zaili Yang, Ramin Riahi and Jin Wang

Abstract Today's maritime environments are characterized by high uncertainties due to the diverse risks associated with seaports' operations. Although much effort has been made in developing new methods to prevent incidents in seaports, yet security breaches still occur largely because of the complex interactions of the multiplicity of stakeholders involved in their operations such as the terminal operators, seafarers and law enforcement agencies which often lead to the disruption of their systems. Experience has shown that complexities in seaports' systems do not allow for flexible response to security incidents. In order to address the numerous risks associated with seaport operations, a subjective security risk analysis is developed to enhance the security of the system. In the analysis process, security systems/measures are identified and described, their performance is measured based on three risk parameters; the likelihood of threat (l), vulnerability of the system (v) and consequence or impact (c) resulting from a successful threat exercise of a flaw in the system. A Fuzzy Analytical Hierarchy Process (FAHP) is utilized to analyse the complex structure of the system and determine the weights of security systems/measures while evidential reasoning (ER) is used to synthesize the risk analysis. It is envisaged that the proposed approach could provide analysts with a flexible tool to understand the importance of developing robust resilience strategies that aim at enhancing seaport security in a systematic manner.

Keywords Maritime transport · Maritime security · Maritime risk · Performance effectiveness · Analytical hierarchy process · Evidential reasoning

A. John · Z. Yang (✉) · R. Riahi · J. Wang
Liverpool Logistics, Offshore and Marine Research Institute,
Liverpool John Moores University, Liverpool, UK
e-mail: Z.yang@ljmu.ac.uk

C. Konstantopoulos and G. Pantziou (eds.), *Modeling, Computing and Data Handling Methodologies for Maritime Transportation*, Intelligent Systems Reference Library 131, DOI 10.1007/978-3-319-61801-2_6

6.1 Introduction

Critical maritime infrastructure (CMI) systems, which are defined as ports, waterways, vessels and their intermodal connections, are the backbone of world economic development due to the enormous roles they performed in enhancing smooth flows of cargoes around the globe. Over the past few decades, these systems have arguably been vulnerable to a wide variety of crimes ranging from theft to direct attacks by terrorists who utilise the systems as a conduit for shipment of Weapons of Mass Destruction (WMD), hostile operatives and contraband.

Given the dynamic risks and the complex operational environment of maritime operations, the threats affecting the systems and their economic realities are well researched and documented [19]. However, the implemented International Shipboard and Port Facility Security (ISPS) Code requires security risk assessment for various ship and port facility security plans to be designed in such a manner that it can prevent and detect security flaws within an international framework, enabling collection and exchange of security information, providing a methodology for assessing security, and ensuring that adequate measures are in place and roles and responsibilities of port facility security officers are designated [30].

Although the Code does show some level of effectiveness in enhancing maritime security it does not prescribe a generally accepted framework for maritime security assessment [52, 53]. An obvious problem is that part B of the Code leaves analysts to choose and define their suitable methodology for individual maritime security assessment. Other security measures implemented to optimize CMI systems' operations include the Container Security Initiative (CSI) and the Customs-Trade Partnership Against Terrorism Initiative (C-TPAT) [5, 7, 36, 37].

Several optional maritime security initiatives were developed in response to the 9/11 attacks. The US Department of Homeland Security led the federal effort in developing a comprehensive national maritime security to address all maritime related threats [13]; the offshore and on-shore security assessment tools [14], maritime security risk analysis model (MSRAM) and MSRAM-PLUS [1] have been presented to enrich the insufficient literature of maritime security. However, the lack of data to analyse the complex structure of maritime operations in order to address security scenarios proactively, makes it difficult to adapt the conventional security risk assessment methods.

The purpose of a collaborative modelling of security systems/measures is to prioritize the high-level risks or flaws within the system in order to invest appropriate strategies that aim at enhancing the performances of security systems/measures. Realising such an objective requires other contextual factors to be considered i.e. environmental, economic, technical and organizational factors. These factors can be defined as multiple uncertain decision attributes in analysing a complex maritime security problem.

Yang et al. [52] developed a security assessment framework using fuzzy evidential reasoning approaches to deal with the incompleteness of objective data. Although the method was accepted due to its novelty in enriching maritime security assessment literature, it has lately been criticised in a number of areas due to its complex mathematical algorithm. Furthermore, the approach did not successfully address maritime security assessment from a systematic perspective. Therefore, in response to an urgent need for a user-friendly security systems' assessment framework that will guide the quantification of the security risk level in terms of vulnerability, threat and consequence in a systematic manner, the present paper is developed.

This paper introduce the concept of belief degrees to increase the flexibility and confidence of experts in evaluating the performance effectiveness of security systems/measures in order to propose a new security assessment approach for seaport operations. In order to achieve the aforementioned aims, the paper is organised as follows: Sect. 6.2 reviews the literature associated with maritime security. Section 6.3 presents the methodology. Section 6.4 provides a case study to illustrate the applicability of the methodology and Sect. 6.5 presents the conclusion of the study.

6.2 Literature Review

Maritime security encompasses a wide range of attack scenarios due to the complex nature of their operations. Over the last few decades, the focus of security analysts has been on terrorism due to the impact of such an attack. However, the threat of a cyber-attack is increasingly of growing relevance in maritime information, communication and control systems as its occurrence can compromise data confidentiality, integrity and availability. Usually, cyber-attacks targeting seaport systems are in the following areas [27]:

- Sea and land-based systems such as vessel tracking and information system (VTIS), automatic identification system (AIS) or long-range identification and tracking system (LRIT).
- Container terminal operating systems (CTOS).
- Port electronic data interchange (EDI) system for domestic and international trade.

Before the 9/11 attacks, most security incidents associated with port operations were from drugs smuggling and organised crimes. However, the studies on terrorist attacks disrupting port operations and supply chain systems has been growing in the relevant literature in the post 9/11 era [19, 45]. Seaports, as an integral component of critical maritime infrastructure, (CMI) systems face additional security threats due to their close spatial interaction with mega-city agglomerations and seashore tourist attractions [7]. Security experts have shown that some terrorist groups also have varying degrees of maritime expertise and capabilities which can be exploited to wreak havoc on a seaport system through the following means [11, 15, 16, 31–33]:

- Sink a large commercial cargo ship in a major shipping channel, thereby blocking traffic to and from the port.

- Seize control of a large commercial cargo ship and use it as a collision weapon for destroying a bridge or refinery located at the waterfront.
- Use land around the port's system to wreak havoc, possibly on refineries located in industrial port areas and on other port facilities.
- Attack vessels or ports used to supply military operations overseas and interfere with those operations.
- An attack to disrupt the world oil trade and cause large-scale environmental damage.
- An attack on or hijacking of a large ship containing volatile fuel (i.e. Liquefied Natural Gas (LNG) or Liquefied Petroleum Gas (LPG)) and detonation of it to cause in-port explosions.
- Use of commercial cargo containers to smuggle terrorists, nuclear, chemical or biological weapons, components thereof or other dangerous materials into a country.
- Directly target a cruise liner or passenger ferry to cause mass casualties by contaminating the ship's food supply, detonating an improvised explosive device (IED) or ramming the vessel with a fast-approach small attack craft.

6.2.1 Analysis of Maritime Security Systems/Measures

When developing a robust and flexible maritime security programme, it is imperative to consider the vulnerability of the system and its economic realities. In view of the above, seaports need to conduct thorough operational, personnel, informational technology and physical security risk analysis and comprehensive system review in a structured and systematic manner so that threats of high probability can be identified and prioritised for resilience improvement of the system.

As seaports' territories are broad and many stakeholders need to participate actively to achieve maximum benefit, security systems/measures need to be designed and constructed in such a manner in which a realistic response can be produced in the face of attacks, or be built with an adequate safety margin to account for uncertainty. The review of the ISPS Code and the various research [4, 8, 36, 39] revealed the following port security systems/measures for collaborative design and modelling:

- Access control measures.
- Advanced cargo information process systems.
- Screening measures.
- Detection measures.

6.2.1.1 Access Control Measures

Information regarding authorised individuals coming into the port systems needs to be monitored in order to control entry and exit to a port facility and to specific areas within the port systems. The access control system is designed to prevent unauthorised entry. An effective access control system prevents the introduction of harmful devices, materials and components. Additionally, access control systems include guarded entry and exit points, access control rosters, personal recognition, ID cards, badge exchange procedures and personnel escorts for visitors. Due to the complexity of the access control system, cyber access systems have been further developed to optimise the system's operation; they include firewall, password protection, and antivirus software.

6.2.1.2 Detection Measures

Seaport facilities are considered to be critical infrastructure systems and are vulnerable to wide varieties of risks due to their complex structures. As a result, they need to be protected from threats by early detection using sophisticated and robust security systems/measures. The detection capability of a seaport security can be achieved by assessing the following [1]:

- Perimeter intrusion detection device.
- Underwater and sonar cameras.
- Security patrolling and guarding.
- Monitoring systems' devices deployed at strategic locations within the ports systems.

6.2.1.3 Screening System Measures

The use of screening measures in port operations has significantly helped in detecting hazardous, illicit substances and human trafficking in major ports around the world. These developments have enhanced the effectiveness of container cargo shipment and supply chain security with minimum disruption to operations. Three major cargo security initiatives have been developed based on the US Safe Port Act of 2006 to improve maritime and cargo security; these are non-intrusive inspection (NII), radiation scanning and radio frequency identification (RFID) container intrusion detection [15, 17, 29].

6.2.1.4 Advanced Cargo Information Process

In order to enhance the security of ports and the supply chain systems in general due to the perceived threat of terrorists exploiting deficiencies in the system to plan and execute an attack, the Customs and Border Protection agency (CBP) requires advanced cargo information from importers and ocean vessel carriers to be supplied electronically for non-bulk cargo shipments in-bound to the US. The focus of these initiatives is to improve and enhance the CBP's ability to be proactive in cargo security risk assessment and to recognise high-risk shipments going into the US seaports [38]. Pursuant to section 203 of the Safe Port Act of 2006 and section 343(a) of the Trade Act of 2002 as amended by the Maritime Transportation Security Act (MTSA) of 2002, the security of port systems can be evaluated by assessing the following initiatives [38]:

- Importer Security Filing (ISF) (responsible for filing ten (10) data elements).
- Ocean Vessel Carrier Filing (OVCF) (responsible for filing two (2) data elements).

The accuracy of the $(10 + 2)$ data elements and cargo manifest depends on the level of partnership, with various stakeholders involved in international export and import of cargoes. Partnership between the different parties involved in the process facilitates a reduction in security risks and enhances the robustness of the programme and effectiveness of the security systems of a port.

6.3 Methodology

Since the ability of a seaport to maximize its security and ensure continuous operations despite a mishap depends on the availability of the right information at the right time for robust decision making [25], within this paper a generic framework using fuzzy sets theory, Evidential Reasoning (ER) and Fuzzy Analytical Hierarchy Process (FAHP) is proposed. The proposed framework can be implemented as a key part of security assessment during the process of risk evaluation of a seaport operation. The steps involved in the assessment are shown in Fig. 6.1, as summarised below:

1. Identify seaports' security systems/measures.
2. Develop a generic risk model based on 1 above.
3. Determine the weight of each security system/measure.
4. Adopt fuzzy logic and a belief degree concept to calibrate and measure the performance of security systems.
5. Synthesize the assessment results obtained in steps 3 and 4 using ER.
6. Apply the framework to a real case study.
7. Validate the result using appropriate techniques including sensitivity analysis.

6.3.1 Identify Seaports' Security Systems/Measures

As seaports maintain their position on the frontline to facilitate global trade and to ensure a formidable security system, the complex risk environment challenges

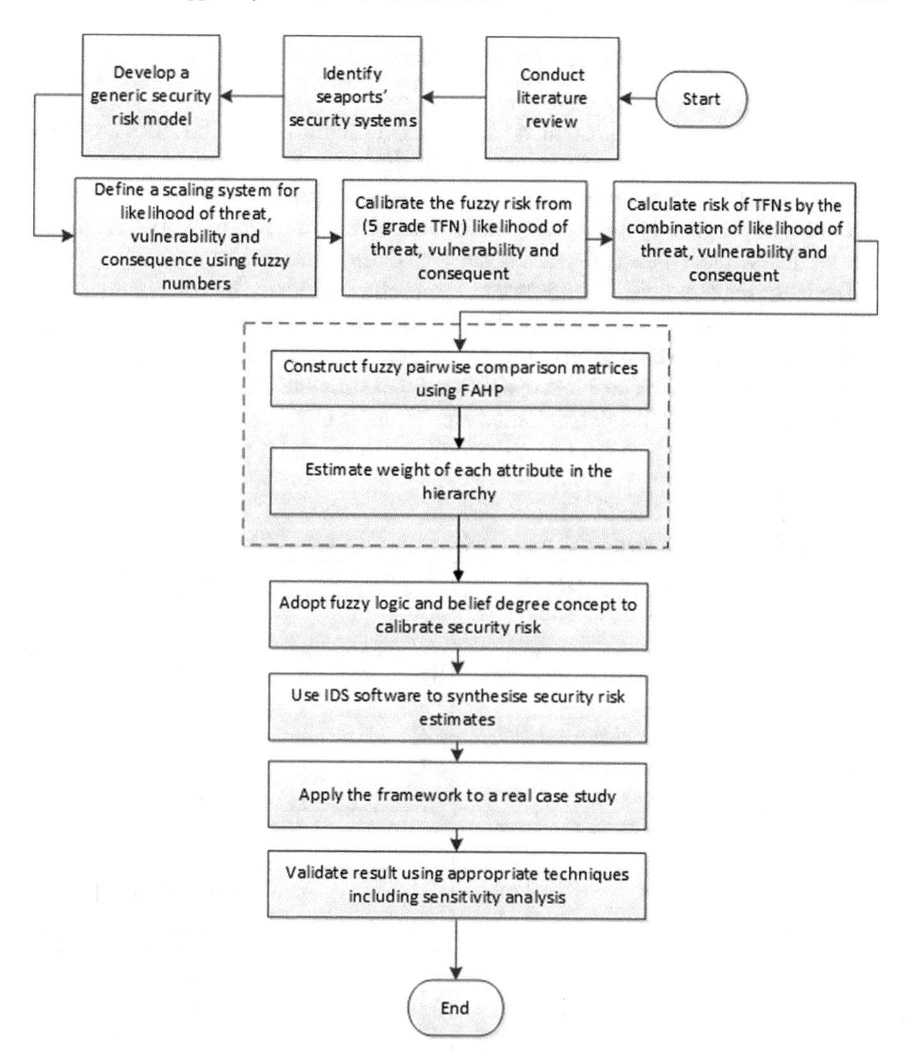

Fig. 6.1 Framework for security risk analysis of a seaport

decision makers with regard to resource allocation among a wide group of countermeasures in order to minimise security risk and enhance the adaptive capacity of the system. The literature review revealed various security systems/measures which can serve as the basis for collaborative modelling and strategic decision making of the system, and which are presented in Table 6.1 [10, 21, 38, 45, 47].

6.3.2 Develop a Generic Security Risk Model

Based on the information presented in Sect. 6.2.1, Table 6.1 and the ISPS Code, an integrated generic security risk model with hierarchical structure is presented in Fig. 6.2. A set of linguistic variables (i.e. Tables 6.2 and 6.3) developed based on literature [28] is used by the security analyst to describe the parameters of the model in order to establish a basis for the modelling of the system.

Decision makers need to understand the problems clearly before attempting to solve them in a transparent manner. This is true when there are many criteria to be considered, each of which in turn consists of several sub-criteria. For this reason, it is imperative to display the problem in a hierarchical structure. Based on the model, the goal of the problem is the performance effectiveness of security systems/measures. In the second level, there are several criteria, each of which has its contribution to

Table 6.1 List of seaport security systems/measures

Abbreviation	Security systems	Abbreviation	Security systems
R1	Access control	R30	Perimeter fencing patrol
R2	Screening measures	R31	Marshalling area patrol
R3	Detection measures	R32	Waterborne area patrol
R4	Advanced cargo information process	R33	Cargo handling area patrol
R5	Personnel security	R34	Access route area patrol
R6	Operational security	R35	Monitoring device for warehousing
R21	Biometric system	R36	Monitoring device for stacking area
R22	Watch-men at quay/vessel interface	R37	Monitoring device for perimeter side
R23	Guards at main security gate	R38	Monitoring device for main gate
R24	Radiation revealing device	R39	Monitoring device for quayside operations area
R25	Gamma/X-ray revealing device	R40	Security patrolling and guarding
R26	Perimeter intrusion detection device	R41	Monitoring system device
R27	Underwater and sonar camera	R29	Ocean vessel carrier filing
R28	Importer security filing		

Table 6.2 Qualitative descriptors for triangular fuzzy numbers

Grade	Vulnerability	Threat likelihood	Consequence severity	Membership functions
1	Very low	Very low	Negligible	(0.0, 0.0, 0.25)
2	Low	Low	Minor	(0.0, 0.25, 0.5)
3	Medium	Medium	Moderate	(0.25, 0.5, 0.75)
4	High	High	Serious	(0.5, 0.75, 1.0)
5	Very high	Very high	Catastrophic	(0.75, 1.0, 1.0)

Table 6.3 Security risk evaluation membership function

Evaluation of security systems/measures	Grades	Membership function
Very low	1	(0.0, 0.0, 0.25)
Low	2	(0.0, 0.25, 0.5)
Medium	3	(0.25, 0.5, 0.75)
High	4	(0.5, 0.75, 1.0)
Very high	5	(0.75, 1.0, 1.0)

measuring the performance of the overall goal, and then some of these criteria are further broken down until the stage where decision makers are able to make informed and practical decisions on resilience of the security systems under high uncertainty.

6.3.3 Determine the Weight of Each Security System/Measure

FAHP is employed in this study to obtain the weight of each attribute in the hierarchy and to synthesise the risks from the bottom to the top level of the hierarchy in a systematic fashion. Compared to the conventional AHP method, which uses crisp values in evaluating the relative importance of each attribute, FAHP uses fuzzy ratios for ease of expert knowledge elicitation.

An advantage of FAHP is its flexibility for integration with different techniques such as evidential reasoning in risk analysis. Therefore, FAHP leads to the generation of weighting factors to represent the primary risk within each category of the model. When determining the weights of attributes, the experts' judgement is in the form of pair-wise comparisons based on an estimation scheme, which lists the intensity of importance using linguistic variables. Each variable has a corresponding triangular fuzzy number that is employed to transfer experts' judgements into a comparisons matrix as presented in Eq. 6.1 [2].

$$\tilde{a}_x = (L, M, U) \tag{6.1}$$

Table 6.4 Weight estimation scheme

Level of importance in qualitative descriptors	Description	Triangular fuzzy numbers (TFNs)
Equal importance	Two attributes contribute equally to the risk of disruption	(1, 1, 2)
Between equal and weak importance	When compromise is needed	(1, 2, 3)
Weak importance	The subjective judgement and experience of experts slightly favour one attribute group over another	(2, 3, 4)
Between weak and strong importance	When compromise is needed	(3, 4, 5)
Strong importance	The subjective judgement and experience of experts strongly favour one attribute group over another	(4, 5, 6)
Between strong and very strong importance	When compromise is needed	(5, 6, 7)
Very strong importance	A given attribute is favoured very strongly over another	(6, 7, 8)
Between very strong and absolute importance	When compromise is needed	(7, 8, 9)
Absolute importance	The evidence favouring one attribute group over another is of the highest possible order	(8, 9, 9)

where L, M and U stand for the smallest possible number, the most promising number and the largest possible number that describes a fuzzy event. Table 6.4 shows the linguistic variables for a criterion and its corresponding triangular fuzzy number (TFN), as modified and adopted from [2], it is used in this study for the purpose of weighting factor estimation.

Suppose there are m experts or decision makers with equal weights, the elements in a fuzzy pair-wise comparison matrix can be modelled as follows:

$$\tilde{a}_{i,j} = \left(\frac{1}{m}\right) \otimes \left(e_{i,j}^{1} \oplus e_{i,j}^{2} \oplus \cdots e_{i.j}^{k} \cdots \oplus e_{i,j}^{m}\right) \tag{6.2}$$

$$\tilde{a}_{j,i} = \frac{1}{\tilde{a}_{i,j}} \tag{6.3}$$

where $\tilde{a}_{i,j}$ is the relative importance by comparing event i and j while $e_{i.j}^{k}$ represents the k_{th} expert judgement in TFN format and $\otimes$ is fuzzy multiplication operation. For a $n \times n$ fuzzy pair-wise comparison matrix, $\tilde{A}$ can be obtained as follows:

$$\tilde{A} = \begin{pmatrix} \tilde{a}_{1,1} & \tilde{a}_{1,2} & \dots & \tilde{a}_{1,n} \\ \tilde{a}_{2,1} & \tilde{a}_{2,2} & \dots & \tilde{a}_{2,n} \\ \vdots & \vdots & \tilde{a}_{i,j} & \vdots \\ \tilde{a}_{n,1} & \tilde{a}_{n,2} & \dots & \tilde{a}_{n,n} \end{pmatrix} \tag{6.4}$$

The weight factors of each element in the hierarchy can be computed using the geometric mean technique [9].

$$\tilde{r}_i = \left(\tilde{a}_{i,1} \otimes \tilde{a}_{i,2} \otimes \cdots \otimes \tilde{a}_{i,n}\right)^{1/n} \tag{6.5}$$

$$\tilde{w}_i = \tilde{r}_i \otimes (\tilde{r}_1 \oplus \cdots \oplus \tilde{r}_n)^{-1} \tag{6.6}$$

where $\tilde{a}_{i,n}$ is the fuzzy comparison value of criterion i to criterion n, $\tilde{r}_i$ is the geometric mean of the ith row in the fuzzy pair-wise comparison matrix, and $\tilde{w}_i$ is the fuzzy weight of the ith criterion of a triangular fuzzy number (TFN) indicated by $\tilde{w}_i = (w_i^l, w_i^m, w_i^u)$, while w_i^l, w_i^m and w_i^u are the lower, middle and upper values of the fuzzy weight of the ith criterion respectively.

The geometric mean obtained from the triangular fuzzy weight using Eq. 6.6 needs to be defuzzified into a crisp weight factor using an approach derived by [46].

The defuzzified mean value $DF_{\tilde{w}_i}$ for (w_i^l, w_i^m, w_i^u), can be obtained as follows:

$$DF_{\tilde{w}_i} = \frac{\left(w_i^u - w_i^l\right) + \left(w_i^m - w_i^l\right)}{3 + w_i^l} \tag{6.7}$$

The normalised weight of the ith attribute can be obtained using Eq. 6.8.

$$w_i = \frac{DF_{\tilde{w}_i}}{\sum DF_{\tilde{w}_i}} \tag{6.8}$$

where $\sum DF_{\tilde{w}_i}$ represents the sum of the defuzzified mean values of all the rows in the comparison matrix.

In order to control and ensure accuracy in the result of the method, the consistency ratio for each of the matrices needs to be analysed. The consistency ratio (CR) is used to estimate the consistency of the pair-wise comparisons as follows:

$$CR = CI/RI$$

$$CI = \frac{\lambda_{max} - n}{n - 1}$$

$$\lambda_{max} = \frac{\sum_{j=1}^{n} \frac{\sum_{k=1}^{n} w_k a_{jk}}{w_j}}{n} \tag{6.9}$$

where *CI* stands for consistency index, *RI* stands for average random index [41], n stands for matrix order, and λ_{max} stands for maximum weight value of the n-*by*-n comparison matrix. When *CR* less than 0.10, the comparisons are acceptable; otherwise, they are not acceptable and should be revised in order to obtain a consistent opinion [35].

6.3.4 Adopt Fuzzy Logic and a Belief Degree Concept to Calibrate the Performance of Security Systems

The development of an effective security risk model is an important mechanism in the quest to accurately assess the risks associated with the flaws within a seaport system. In the security context, risk assessment focuses on assessing the likelihood of attack, vulnerability of the system, and consequence given the success of a variety of threat scenarios. Modern seaports require an integrated security risk-based model that combines the elements of vulnerability assessments, threat and consequence scenarios in a dynamic environment. The focus is on cargo, people and information process flows when determining risks and performance of security systems, taking into account ports' security forces. Based on the identified security systems/measures presented in Table 6.1, a generic seaport security model is developed and presented in Fig. 6.2. The security systems/measures are selected based on robust literature review and extensive discussions with experts who revealed them as the most significant systems often used in seaports' operation. The model is a tool for decision making to help port operators optimise both security and resource utilisation.

6.3.4.1 Security Risks' Representation Using Fuzzy Set Theory

Risks in CMI systems in which a seaport is an integral component can be defined as a set of triplets containing an initiating event (i.e. the probability of an event occurring), vulnerability of the system, which comprises a set of system or target weaknesses that can be exploited by an adversary to achieve a given degree of loss or harm, and the consequence of the event given its occurrence [23].

The occurrence likelihood of an initiating event is measured in terms of its frequency or probability and is largely driven by adversarial (e.g. terrorist) capability to exploit any flaws in security systems/measures. The consequent severity indicates the impact of the losses following the occurrence of undesired event due to flaws within the system's operations. Vulnerability is measured in terms of the effectiveness of the port's existing security programme and response capabilities which can be exploited by the adversary to achieve a given degree of success in terms of loss

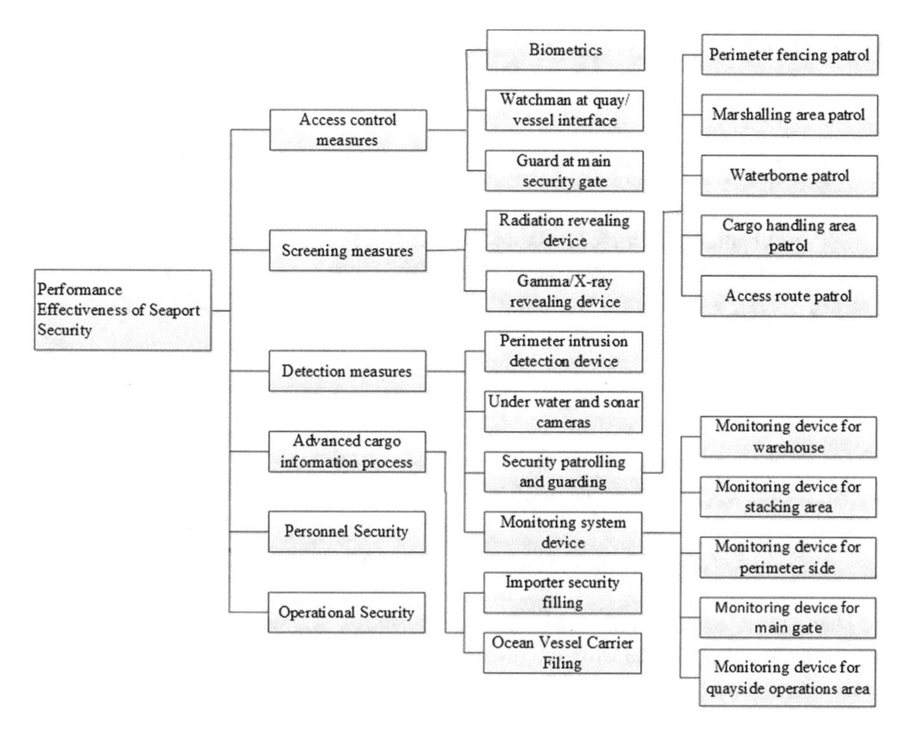

Fig. 6.2 Generic seaport systems' security risks model

or harm. Though there are some overlaps in the description of the models' attributes and the risk parameters, the main issue or content are largely independent which can enable their aggregation and synthesis via the ER software for a flexible decision-making process. Based on the above analogy, risk can be represented mathematically as:

$$Risk\ r = Threat\ \ likelihood\ (l) \otimes\ Vulnerability\ (v) \otimes\ Consequence\ (c) \quad (6.10)$$

Equation 6.10 is applicable to all categories of risk associated with CMI systems and can be used to determine the risk level of each security system/measure in order to evaluate its performance. Therefore, this will provide the decision makers with an opportunity to adapt appropriate strategies that are robust yet flexible enough to reduce the probability of an attack through deterrence and countermeasures. Security risks can be reduced by improving the performance of the security systems/measures. The literature review indicated that a common expression for risk presented in Eq. 6.10 is seen as the philosophical basis for security risk assessment methodologies of many maritime critical systems [26]. Therefore, Eq. 6.10 is assumed independent and represented by fuzzy numbers for a flexible modelling of the system as indicated by Mokhtari et al. [24], UNISYS [47], Sadiq et al. [42],

Mcgill et al. [23], Anoop et al. [3] and Moteff [26], Eq. 6.10 can be written as:

$$\tilde{r} = \tilde{l} \otimes \tilde{v} \otimes \tilde{c} \tag{6.11}$$

where $\tilde{r}$, $\tilde{l}$, $\tilde{v}$ and $\tilde{c}$ stand for fuzzy risk, likelihood of threats, vulnerability and consequence respectively.

The implication for Eq. 6.11 is that the risk levels of every system under a fuzzy environment can be analysed as the product of the three fuzzy numbers represented as $\tilde{l} = (x_l, y_l, z_l)$, $\tilde{v} = (x_v, y_v, z_v)$ and $\tilde{c} = (x_c, y_c, z_c)$. Based on Eq. 6.11, these numbers can be expressed as follows:

$$FTN_{lvc} = FTN_l \otimes FTN_v \otimes FTN_c \tag{6.12}$$

where *FTN* represents fuzzy triangular numbers. *FTN* is used because of its computational simplicity and the ease with which it can be applied during the calculation process.

$$FTN_{lvc} = (x_l \otimes x_v \otimes x_c, y_l \otimes y_v \otimes y_c, z_l \otimes z_v \otimes z_c) \tag{6.13}$$

Due to the dynamic nature of maritime business, seaport systems require good scientific and engineering knowledge on diverse issues relating to the systems' operations. The generic security model considers operational factors such as cargo, people and information process flows when determining risks and action plans, while taking into account potential adversaries and the port's security forces, as well as threats and vulnerabilities including decisions on security resource utilisation.

6.3.4.2 Linguistic Variables for Security Risk Parameters

Because security risk is a fuzzy problem that is uncertain and imprecise, it is usually challenging to quantify it due to the fact that potential security flaws or threats occur infrequently and their closed interval range can have an assumed value of 0 and 1. A practical and efficient way to express security levels in a port is to use qualitative descriptors, particularly from the port facilities security officers (PFSOs), company security officers (CSOs) and security intelligent experts.

The likelihood of threats and vulnerability can be assessed using such terms as Very Low, Low, Medium, High and Very High while the consequence or impact can be assessed as Negligible, Minor, Moderate, Critical and Catastrophic, as represented in Tables 6.5, 6.6 and 6.7 respectively. These subjective variables can further be defined in terms of their membership functions with a curve that defines how each point in the input space can be mapped into a membership value between 1 and 0. As presented in Riahi et al. [39], the most commonly used membership functions are the triangular and trapezoidal; this research uses the five scale method, adapted and modified from Ngai and Wat [28] to represent the l, v and c levels of security risk as shown in Fig. 6.3 with uniform distribution of linguistic variables.

Table 6.5 Definition of fuzzy linguistic variables for likelihood assessment

Risk factors	Linguistic terms	Grade	Meaning
Likelihood of threat (Probability of false information or accidentally triggering or intentionally exploiting a specific vulnerability to cause damage, harm or loss)	Very low (VL)	1	There is likelihood of false information or threat against the infrastructure
	Low (L)	2	There is little evidence of false information or threat against the infrastructure
	Medium (M)	3	There is likely evidence of false information or threat against the infrastructure
	High (H)	4	There is relatively reliable evidence of false information or threat against the infrastructure
	Very high (VH)	5	There is reliable evidence of false information or threat against the infrastructure

6.3.4.3 Application of the Belief Degrees in Security Risk Assessment

Several decision problems in engineering and management involve multiple attributes presented in different quantitative and qualitative forms. Furthermore, a decision may not be correctly made without fully taking into consideration all the elements in question [6, 18, 40, 41, 44, 50]. There is a close relationship between complexity and uncertainty and it is said that when complexity increases, certainty decreases. Therefore, there is a need to develop a robust yet flexible technique for dealing with Multiple Attribute Decision Analysis (MADA) problems under uncertainty in a manner that is reasonable, reliable, and transparent [51]. Eventually, in order to measure the performance effectiveness of each security system/measure, it is necessary to transform the fuzzy ratings of all parameters into belief structures with the same set of evaluation grades [20].

A belief degree generally represents the strength to which an answer is believed to be true, and it must be equal to or less than 100% or able to be described as the degree of expectation that, given an alternative, will yield an anticipated outcome on

Table 6.6 Definition of fuzzy linguistic variables for vulnerability assessment

Risk factors	Linguistic ratings	Grade	Meaning
Vulnerability (A flaw or weakness in the seaport system's security procedures, design, implementation or internal controls that could be exploited and result in a security breach or a violation of the system's security policy)	Very low (VL)	1	The threat source has very little motivation or capability of carrying out attacks. Controls are in place to prevent, or significantly impede, the vulnerability from being exploited
	Low (L)	2	The threat source has little motivation or capability of carrying out attacks. Controls are in place to prevent, or significantly impede, the vulnerability from being exploited
	Medium (M)	3	The threat source is motivated and capable of carrying out attacks. Controls to prevent the vulnerability from being exercised are moderately ineffective
	High (H)	4	The threat source is fairly motivated and capable of carrying out attacks. Controls to prevent the vulnerability from being exploited are ineffective
	Very high (VH)	5	The threat source is highly motivated and sufficiently capable of carrying out attacks. No controls are available to prevent the vulnerability from being exploited

Table 6.7 Definition of fuzzy linguistic variables for consequence assessment

Risk factors	Linguistic ratings	Grade	Meaning
Impact (Consequence)	Negligible (N)	1	Impact has no effect on the infrastructure, it may not be noticed
	Minor (Mi)	2	Impact would cause slight annoyance to the personnel but not result in infrastructure deterioration
	Moderate (Mo)	3	Impact would cause a high degree of operation dissatisfaction or result in noticeable but slight infrastructure deterioration
	Serious (S)	4	Impact would cause significant deterioration in infrastructure system performance and/or lead to injuries
	Catastrophic (Ca)	5	Impact would seriously affect the ability to complete a task or would cause damage to infrastructure and environment and/or lead to injuries or even death

a particular criterion. The use of individual belief degrees depends on an analyst's expertise, knowledge of the subject matter and experience regarding the operation of the system. The justification for the use of belief degrees is as a result of the fact that human decision making involves ambiguity, uncertainty and imprecision; individuals make judgements in probabilistic terms with the help of their knowledge.

In risk analysis, one realistic way to analyse a security risk with incomplete objective data is to employ a fuzzy *IF-THEN* rule built from human understanding, where premise and conclusions contain the linguistic variables used to describe risk parameters [52]. Having identified the maritime security risk parameters and their corresponding linguistic variables, fuzzy *IF-THEN* rules with a belief structure can be constructed to model a security risk assessment scenario. For example, an *IF-THEN* rule can be developed as follows:

If Threat Likelihood is "Medium", system Vulnerability is "High" and Impact or consequent severity is "Serious", then security risk is "High". Accordingly, when

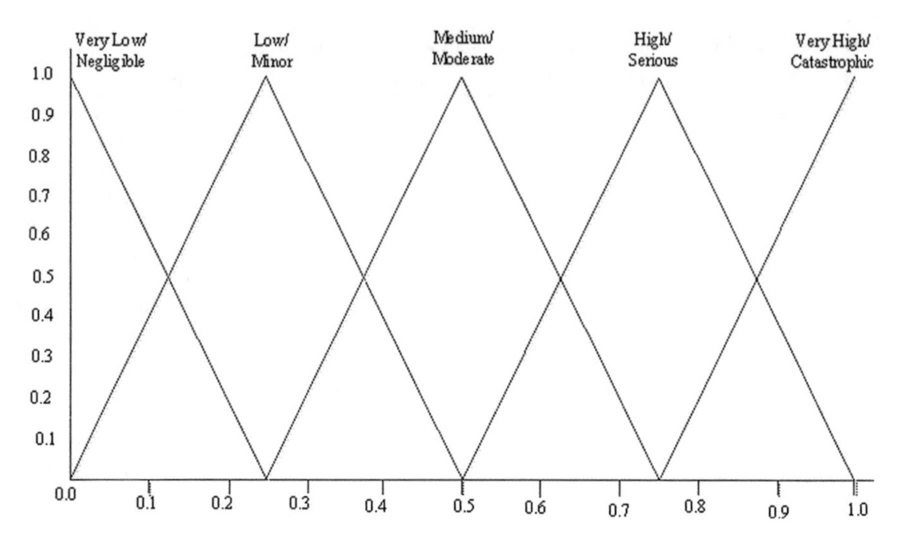

Fig. 6.3 Fuzzy triangular membership functions

measuring the risk level of each security attribute it is necessary to convert the fuzzy rating of the parameters into a belief structure with the same set of evaluation grades [20]. The evaluation of security risk attributes by each of the risk parameters can be explained by the following expression:

$$Z = [Z_1, Z_2, Z_3, Z_4, Z_5] = \{\text{Very Low, Low, Medium, High, Very High}\} \quad (6.14)$$

The estimation of security risk level obtainable in Eqs. 6.10–6.13 can be converted using the five steps presented in Table 6.8. The obtained result (i.e. FTN_{lvc}) can be

Table 6.8 Five steps of converting FTN_{lvc} into fuzzy risk Z

Step 1	Map the calculated FTN_{lvc} over FTN_r (i.e. 5 grades defined over the universe of discourse of risk (VL, L, M, H and VH))
Step 2	Determine the point where the newly mapped FTN_{lvc} intersects each linguistic term of the FTN_r
Step 3	Use a maximum figure if FTN_{lvc} and a linguistic term of FTN_r intersect at more than one point
Step 4	Establish a set of intersecting points that defines the 5 non-normalised grades in the form of fuzzy sets
Step 5	Normalise the Z_r (5 non-normalised grades) to obtain Z (5 normalised grades) which is known as the belief structure

converted into fuzzy risk Z (i.e. the normalised fuzzy set) [20, 42]. The conversion of FTN_{lvc} into Z is carried out to simplify the computational analysis and present the risk's level in a unified space of discourse which can subsequently be used as input data into the IDS software for aggregation and ranking in a systematic fashion.

6.3.5 Application of Evidential Reasoning Algorithm (ER) for Security Risk Synthesis

The theory of evidence was first generated by Dempster [12] and further developed by Shafer [43]. It is often referred to as Dempster–Shafer theory of evidence or D–S theory. The D–S theory was originally used for information aggregation in expert systems as an approximate reasoning tool [22]. Subsequently it has been used in decision making under uncertainty [49]. Due to the ever-changing environment and the multiple criteria decision making problems having a degree of uncertainty, the ER algorithm was developed. The ER approach can be elucidated as follows [51]:

Let "R" represent the set of the five risk expressions and be synthesised by two subsets R_1 and R_2 from two different assessors. Then, for example, R, R_1 and R_2 can separately be expressed by:

$R = \{\beta^1$ "*Very Low*", β^2 "*Low*", β^3 "*Medium*", β^4 "*High*", β^5 "*Very High*"$\}$
$R_1 = \{\beta_1^1$ "*Very Low*", β_1^2 "*Low*", β_1^3 "*Medium*", β_1^4 "*High*", β_1^5 "*Very High*"$\}$
$R_2 = \{\beta_2^1$ "*Very Low*", β_2^2 "*Low*", β_2^3 "*Medium*", β_2^4 "*High*", β_2^5 "*Very High*"$\}$

where "*Very Low*", "*Low*", "*Medium*", "*High*" and "*Very High*" (the risk expression) are associated with their corresponding degrees of belief. Suppose the normalised relative weights of two assessors in the risk evaluation process are given as ω_1 and ω_2 ($\omega_1 + \omega_2 = 1$). ω_1 and ω_2 can be estimated by using an AHP technique. Suppose M_1^m and M_2^m ($m = 1, 2, 3, 4$ or 5) are individual degrees to which the subsets R_1 and R_2 support the hypothesis that the risk evaluation is confirmed to the five risk expressions. Then, M_1^m and M_2^m are obtained as follows:

$$\begin{aligned} M_1^m &= \omega_1 \times \beta_1^m \\ M_2^m &= \omega_2 \times \beta_2^m \end{aligned} \tag{6.15}$$

where $m = 1, 2, 3, 4, 5$.

Suppose H_1 and H_2 are the individual remaining belief values unassigned for M_1^m and M_2^m ($m = 1, 2, 3, 4, 5$). Then, H_1 and H_2 are expressed as follows [51]:

$$\begin{aligned} H_1 &= \bar{H}_1 + \tilde{H}_1 \\ H_2 &= \bar{H}_2 + \tilde{H}_2 \end{aligned} \tag{6.16}$$

where $\bar{H}_n (n = 1 \text{ or } 2)$, representing the degree to which the other assessor can play a role in the assessment, and $\tilde{H}_n (n = 1 \text{ or } 2)$, is caused by the possible incompleteness in the subsets R_1 and R_2. $\bar{H}_n (n = 1 \text{ or } 2)$ and $\tilde{H}_n (n = 1 \text{ or } 2)$, are described as follows:

$$\begin{aligned} \bar{H}_1 &= 1 - \omega_1 = \omega_2 \\ \bar{H}_2 &= 1 - \omega_2 = \omega_1 \\ \tilde{H}_1 &= \omega_1 \left(1 - \sum_{m=1}^{4} \beta_1^m \right) \\ \tilde{H}_2 &= \omega_2 \left(1 - \sum_{m=1}^{4} \beta_2^m \right) \end{aligned} \tag{6.17}$$

Suppose $\beta^{m'}$ $(m = 1, 2, 3, 4 \text{ or } 5)$ represents the non-normalised degree to which the risk evaluation is confirmed to each of the five risk expressions as a result of the synthesis of the judgments produced by assessors 1 and 2. Suppose $H_U{}'$ represents the non-normalised remaining belief unassigned after the commitment of belief to the five risk expressions because of the synthesis of the judgments produced by assessors 1 and 2. The ER algorithm is stated as follows [51]:

$$\begin{aligned} \beta^{m'} &= K \left(M_1^m M_2^m + M_1^m H_2 + M_2^m H_1 \right) \\ \bar{H}_U{}' &= K \left(\bar{H}_1 \bar{H}_2 \right) \\ \tilde{H}_U{}' &= K \left(\tilde{H}_1 \tilde{H}_2 + \tilde{H}_1 \bar{H}_2 + \tilde{H}_2 \bar{H}_1 \right) \\ K &= \left[1 - \sum_{T=1}^{4} \sum_{\substack{R=1 \\ R \neq T}}^{4} M_1^T M_2^R \right]^{-1} \end{aligned} \tag{6.18}$$

After the above aggregation, the combined degrees of belief are generated by assigning $\bar{H}_U{}'$ back to the five risk expressions using the following normalization process [51]:

$$\begin{aligned} \beta^m &= \beta^{m'} / 1 - \bar{H}_U{}' (m = 1, 2, 3, 4) \\ H_U &= \tilde{H}_U{}' / 1 - \bar{H}_U{}' \end{aligned} \tag{6.19}$$

where H_U is the unassigned degree of belief representing the extent of incompleteness in the overall assessment. The above gives the process of combining two subsets. If three subsets are required to be combined, the result obtained from the combination of any two subsets can be further synthesized with the third one using the above algorithm. In a similar way, the judgements of multiple assessors or the risk evaluations of lower level criteria in the chain systems (i.e. components or subsystems) can also be combined.

6.3.6 *Perform Sensitivity Analysis*

There are numerous methods for validating a generic knowledge-based system under uncertainty, but the most commonly used validation techniques are field test, subsystem validation, informal validation and sensitivity analysis. Sensitivity analysis has been found to be a reliable technique for validating novel approaches under high levels of uncertainty [34].

This study seeks to examine the sensitivity of the security systems' performances to individual elements that make up the model. Due to the lack of visibility regarding port security issues, the lack of precise data and the novelty of this model, it has not been possible to find any established means of benchmarking the results; this analysis uses incremental processes through conducting several industrial case studies to validate the results. Thus, the developed model can then be refined and applied in the industry to improve the resilience of maritime security systems/measures.

In light of the above, the generic model has been partially validated using sensitivity analysis. The aim of sensitivity analysis is to test the sensitivity of the model; it refers to how sensitive the conclusions are to a minor change in the inputs. The change may be the variation of the model's parameters or changes in the belief degrees assigned to the linguistic variables used to describe them. If the methodology is sound and its inference reasoning is logical, then the sensitivity analysis must at least follow the following axioms [52]:

Axiom 1: A slight decrement or increment in the belief degree of each security system/measure at the bottom level of the hierarchy should result in the effect of the relative decrease or increase in the model output.

Axiom 2: If K and Q, where $(Q > K)$ criteria from all the bottom level criteria are selected and the degree of belief associated with the highest preference linguistic term of each of such K and Q criteria is decreased by the same amount (i.e. simultaneously the degree of belief associated with the lowest-preference linguistic terms of each of such K and Q criteria is increased by the same amount) and the model's output data are evaluated as A_K and A_Q respectively, accordingly, A_Q should be less than A_K.

6.4 Test Case

Based on the generic model presented in Fig. 6.2 and available information in Sect. 6.2.1, a case study is presented to illustrate the applicability of the methodology. The aim of the case study is to provide the decision makers with a simplified means of modelling security scenarios under a fuzzy environment especially during port security risks' auditing, marine terminal security risks' evaluation, and port/terminal redevelopment security systems' planning, construction and implementation.

6.4.1 Identify Seaports' Security Systems/Measures

This phase of the analysis involves the identification of the seaport security systems/measures through a robust literature review and brainstorming session conducted with security analysts using a structured and systematic approach. Accordingly, the security systems/measures are identified and presented in Table 6.1.

6.4.2 Develop a Generic Security Systems/Measures' Risk Model

The identified security systems/measures based on extensive interviews with experts and surveys are represented in a hierarchical structure (see Fig. 6.2) in order to provide insight to security analysts (PFSOs) for a collaborative design and modelling of the system.

6.4.3 Weight Estimation of Security Risks' Attributes

To estimate the relative influence of each sub-criterion to its associated upper-level criterion, it is necessary to assign weight to each sub-criterion. FAHP is employed to achieve the estimation process. In line with the modelling approach presented in An et al. [2], five linguistic variables, presented in Table 6.3 are used for this analysis. The rating of each security system/measure is conducted by a group of three experienced security analysts from industry and academia assigned with equal weights, and the pair-wise comparison is achieved by sets of linguistic variables shown in Table 6.4.

As an example, let R_{28}, and R_{29} represent importer security filing and ocean vessel carrier filing respectively. In view of the above, three experts made these comparisons of R_{28} with R_{29}. The first expert's estimation was "weak importance" which corresponds to triangular fuzzy number (TFN) $(2, 3, 4)$. The second expert's judgement was "strong importance" and the TFN is $(4, 5, 6)$. The third expert's evaluation was between "strong and very strong importance" which corresponds to TFN of $(5, 6, 7)$. By using Eqs. 6.2 and 6.3, the elements in $\tilde{a}_{12}$ and $\tilde{a}_{21}$ can be obtained as follows:

$$\tilde{a}_{1,2} = \frac{1}{3}((2,3,4) \oplus (4,5,6) \oplus (5,6,7)) = (3.6, 4.6, 5.6)$$

$$\tilde{a}_{2,1} = \frac{1}{\tilde{a}_{1,2}} = (0.176, 0.217, 0.270)$$

A 2×2 fuzzy pairwise comparison matrix $\tilde{\delta}$ can be constructed as follows:

$$\tilde{\delta} = \begin{array}{c} \\ a_{11} \\ a_{21} \end{array} \begin{array}{c} \begin{array}{cc} a_{11} & a_{12} \end{array} \\ \begin{bmatrix} (1,\ 1,\ 1) & (3.6,\ 4.6,\ 5.6) \\ (0.18,\ 0.22,\ 0.27) & (1,\ 1,\ 1) \end{bmatrix} \end{array}$$

Each weight of the attribute in the hierarchy can be calculated by using Eqs. 6.5 and 6.6. For example, $\tilde{w}_{11}$ is calculated as follows:

$$\tilde{r}_1 = ((1,1,1) \otimes (3.6,4.6,5.6))^{\frac{1}{2}} = (1 \times 3.6)^{\frac{1}{2}}, (1 \times 4.6)^{\frac{1}{2}}, (1 \times 5.6)^{\frac{1}{2}}$$
$$= (1.897, 2.145, 2.367).$$

In a similar fashion, $\tilde{r}_2$ is calculated accordingly.

$$\tilde{r}_2 = (0.42, 0.466, 0.52)$$

$$\tilde{w}_{11} = \frac{\tilde{r}_1}{\tilde{r}_1 + \tilde{r}_2} = \frac{1.897, 2.145, 2.367}{2.317, 2.611, 2.886} = (0.657, 0.822, 1.021)$$

$\tilde{w}_{11}$ can be converted into a crisp value by using Eq. 6.7 as follows:

$$d\tilde{w}_{11} = \frac{(1.021 - 0.657) + (0.822 - 0.657)}{3 + 0.657} = 0.145$$

In a similar fashion, defuzzified weight $d\tilde{w}_{21}$ is found as 0.035. The normalised weight of w_{11} can be calculated using Eq. 6.8 as follows:

$$w_{11} = \frac{0.145}{0.145 + 0.035} = 0.81.$$

w_{21} is obtained as 0.19.

It is interesting to note that the weights of other security attributes in the hierarchy can be calculated by repeating the above process, and the results obtained are presented in Table 6.9.

Table 6.9 Weight calculations' result for security risks attributes

S/No	Attributes	Description	Weights
1	R1	Access control measures	0.526
2	R2	Screening measures	0.014
3	R3	Detection measures	0.193
4	R4	Advanced cargo information process	0.172
5	R5	Personnel security	0.039
6	R6	Operational security	0.056
7	R21	Biometrics	0.100
9	R22	Watchman at quay/vessel interface	0.200
10	R23	Guard at main security gate	0.700
11	R24	Radiation detection measures	0.435
12	R25	Gamma/X-ray detection measures	0.565
13	R26	Perimeter intrusion detection device	0.590
14	R27	Underwater and sonar cameras	0.220
15	R40	Security patrolling and guarding	0.020
16	R41	Monitoring system devices	0.170
17	R28	Importer security filing	0.810
18	R29	Ocean vessel carrier filing	0.190
19	R30	Perimeter fencing patrol	0.200
20	R31	Marshalling area patrol	0.200
21	R32	Waterborne area patrol	0.200
22	R33	Cargo area patrol	0.200
23	R34	Access route area patrol	0.200
24	R35	Monitoring device for warehousing area	0.200
25	R36	Monitoring device for stacking area	0.200
26	R37	Monitoring device for perimeter side	0.200
27	R38	Monitoring device for main gate	0.200
28	R39	Monitoring device for quayside operations area	0.200

6.4.4 Assessment of the Seaports' Security Systems/Measures Using a Belief Degree Concept

An evaluation sheet was used by the security experts based on the model in Fig. 6.2 to facilitate the analysis of security risks under a fuzzy environment. The experts prioritised the security systems/measures in terms of likelihood of threats, vulnerability and consequence.

As an example, the ratings of the security experts on "ocean vessel carrier filing" are made as (4, 4, 3), representing the likelihood of information misrepresentation or false information provided (0.5, 0.75, 1), vulnerability (i.e. probability of a flaw in the system) (0.5, 0.75, 1) and consequence or impact (0.25, 0.5, 0.75) respectively.

Based on Eq. 6.12, the security risk of the attributes under fuzzy environments is computed as:

$$FTN_{lvc} = 0.06, 0.28, 0.75$$

The obtained risk result FTN_{lvc} is mapped over FTN_r (i.e. 5 grades defined over the universe of discourse of risk (VL, L, M, H and VH)) as shown in Fig. 6.4.

Based on Fig. 6.4, the point where the newly mapped FTN_{lvc} intersects each linguistic term of FTN_r are circled, and maximum values are used at points where FTN_{lvc} and a linguistic term of FTN_r intersect at more than one points, and the corresponding results are presented in Table 6.13. Z_P (i.e. the intersecting points) is normalised to obtained Z. These steps are demonstrated in Fig. 6.4 and Table 6.10.

The result of the process is presented in Tables 6.11 and 6.12, and Fig. 6.4. In a similar way, the experts' ratings for the other factors are analysed and the results obtained are shown in Tables 6.12 and 6.13 respectively.

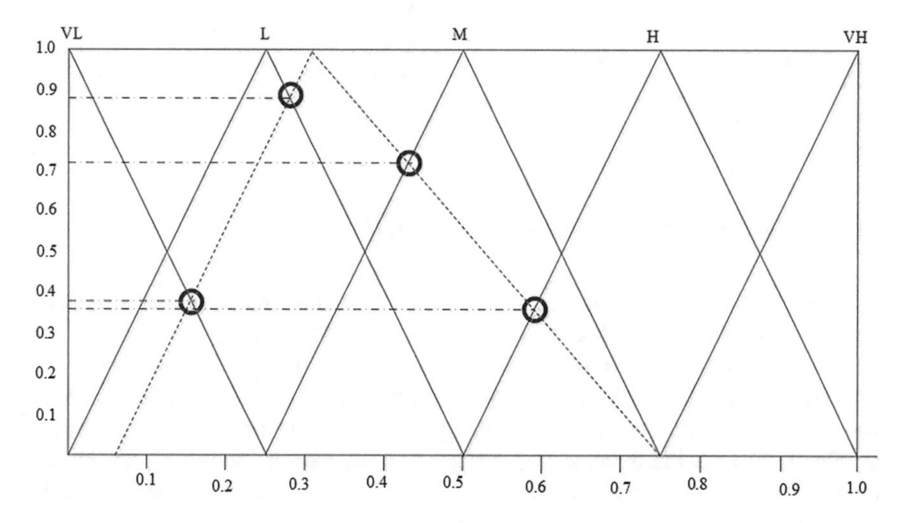

Fig. 6.4 Result of converting FTN_{lvc} to Z_r (5 non-normalised grades)

Table 6.10 Example of converting FTN_{lvc} into Z (belief structures)

FTN_{lvc}	0.06, 0.28, 0.75				
Grade	VL	L	M	H	VH
Z_r	0.39	0.93	0.70	0.35	0
Z	0.16	0.40	0.30	0.14	0

Table 6.11 Obtained results based on the procedures presented in table

Security risks	L	V	C	FTN_l	FTN_v	FTN_c	FTN_{lvc}
R21	4	3	4	(0.50, 0.75, 1.00)	(0.25, 0.50, 0.75)	(0.50, 0.75, 1.00)	(0.06, 0.28, 0.75)
R22	3	4	5	(0.25, 0.50, 0.75)	(0.50, 0.75, 1.00)	(0.75, 1.00, 1.00)	(0.10, 0.38, 0.75)
R23	5	4	5	(0.75, 1.00, 1.00)	(0.50, 0.75, 1.00)	(0.75, 1.00, 1.00)	(0.28, 0.75, 1.00)
R24	3	4	3	(0.25, 0.50, 0.75)	(0.5, 0.75, 1.00)	(0.25, 0.50, 0.75)	(0.03, 0.19, 0.56)
R25	3	2	4	(0.25, 0.50, 0.75)	(0.00, 0.25, 0.50)	(0.50, 0.75, 1.00)	(0.00, 0.10, 0.38)
R26	3	5	4	(0.25, 0.50, 0.75)	(0.75, 1.00, 1.00)	(0.50, 0.75, 1.00)	(0.10, 0.38, 0.75)
R27	3	3	5	(0.25, 0.50, 0.75)	(0.75, 1.00, 1.00)	(0.75, 1.00, 1.00)	(0.05, 0.25, 0.56)
R28	5	5	3	(0.75, 1.00, 1.00)	(0.75, 1.00, 1.00)	(0.25, 0.50, 0.75)	(0.14, 0.50, 0.75)
R29	4	4	3	(0.50, 0.75, 1.00)	(0.50, 0.75, 1.00)	(0.25, 0.50, 0.75)	(0.06, 0.28, 0.75)
R30	4	3	5	(0.50, 0.75, 1.00)	(0.25, 0.50, 0.50)	(0.75, 1.00, 1.00)	(0.10, 0.38, 0.75)
R31	3	5	4	(0.25, 0.50, 0.75)	(0.75, 1.00, 1.00)	(0.50, 0.75, 1.00)	(0.10, 0.38, 0.75)
R32	4	4	5	(0.50, 0.75, 1.00)	(0.50, 0.75, 1.00)	(0.75, 1.00, 1.00)	(0.19, 0.56, 1.00)
R33	5	3	4	(0.75, 1.00, 1.00)	(0.25, 0.50, 0.75)	(0.50, 0.75, 1.00)	(0.10, 0.38, 0.75)
R34	2	3	4	(0.00, 0.25, 0.50)	(0.25, 0.50, 0.75)	(0.50, 0.75, 1.00)	(0.00, 0.10, 0.38)
R35	4	3	3	(0.50, 0.75, 1.00)	(0.25, 0.50, 0.75)	(0.25, 0.50, 0.75)	(0.03, 0.19, 0.56)
R36	4	5	2	(0.50, 0.75, 1.00)	(0.75, 1.00, 1.00)	(0.00, 0.25, 0.50)	(0.00, 0.19, 0.50)
R37	4	3	5	(0.50, 0.75, 1.00)	(0.25, 0.50, 0.75)	(0.75, 1.00, 1.00)	(0.10, 0.38, 0.75)
R38	3	5	3	(0.25, 0.50, 0.75)	(0.75, 1.00, 1.00)	(0.25, 0.50, 0.50)	(0.05, 0.25, 0.56)
R39	2	4	5	(0.00, 0.25, 0.50)	(0.50, 0.75, 1.00)	(0.75, 0.75, 1.00)	(0.00, 0.19, 0.50)

6.4.5 Synthesis of Results Using the ER

An ER algorithm is then implemented to synthesise the risks from an elementary state to the top level of the hierarchy. Synthesis can be achieved through manual calculation or through the use of a Windows-based and graphically-designed decision support software package (IDS) [51] which provides a flexible and easy to use interface for modelling, simulation and decision making. Ultimately, the normalised results (i.e. Z), also known as the experts' belief degrees presented in Table 6.13 can be fed into the IDS software together with the weights presented in Table 6.9 in order to aggregate the security risks in a comprehensive manner.

Table 6.12 Intersection results and normalised security risks

Security risks	Z_R	Z
R21	(0.39, 0.93, 0.70, 0.35, 0)	(0.16, 0.40, 0.30, 0.14, 0)
R22	(0.28, 0.74, 1.00, 0.43, 0)	(0.10, 0.30, 0.40, 0.20, 0)
R23	(0, 0.30, 0.60, 0.95, 0.55)	(0, 0.10, 0.25, 0.40, 0.25)
R24	(0.52, 0.87,0.48, 0.10, 0.00)	(0.25, 0.45, 0.25, 0.05, 0)
R25	(0.70, 0.70, 0.25, 0, 0)	(0.40, 0.40, 0.20, 0, 0)
R26	(0.28, 0.74, 1.00, 0.43, 0)	(0.10, 0.30, 0.40, 0.20, 0)
R27	(0.45, 0.87, 0.59, 0.10, 0)	(0.20, 0.45, 0.30, 0.05, 0)
R28	(0.19, 0.58, 1.00, 0.50, 0)	(0.10, 0.25, 0.45, 0.20, 0)
R29	(0.39, 0.88, 0.73, 0.36, 0.00)	(0.17, 0.370, 0.30, 0.14, 0.00)
R30	(0.28, 0.74, 1.00, 0.43, 0)	(0.10, 0.30, 0.40, 0.20, 0)
R31	(0.28, 0.74, 1.00, 0.43, 0)	(0.10, 0.30, 0.40, 0.20, 0)
R32	(0.53, 0.95, 0.45, 0, 0)	(0.30, 0.50, 0.20, 0, 0)
R33	(0.28, 0.74, 1.00, 0.43, 0)	(0.10, 0.30, 0.40, 0.20, 0)
R34	(0.70, 0.70, 0.25, 0, 0)	(0.40, 0.40, 0.20, 0, 0)
R35	(0.52, 0.87, 0.48, 0.10, 0)	(0.25, 0.45, 0.25, 0.05, 0)
R36	(0.53, 0.95, 0.45, 0, 0)	(0.30, 0.50, 0.20, 0, 0)
R37	(0.28, 0.74, 1.00, 0.43, 0)	(0.10, 0.30, 0.40, 0.20, 0)
R38	(0.45, 0.87, 0.59, 0.10, 0)	(0.20, 0.45, 0.30, 0.05, 0)
R39	(0.53, 0.95, 0.45, 0, 0)	(0.30, 0.50, 0.20, 0, 0)

Table 6.13 Aggregation results for security risks' attributes

Main criteria	Very low	Low	Medium	High	Very high
Access control measures	0.0973	0.3060	0.3020	0.2294	0.0653
Screening measures	0.2948	0.4498	0.2246	0.0309	0
Detection measures	0.1029	0.3225	0.4028	0.1718	0
Advanced cargo information system	0.0992	0.2556	0.4489	0.1963	0
Personnel security measures	0.2000	0.4500	0.3000	0.0500	0
Operational security measures	0.1600	0.4000	0.3000	0.1400	0
Aggregation results	0.1027	0.3252	0.3404	0.1932	0.0386
Security estimate's crisp value	0.44				

Table 6.13 presents the results of the risk synthesis based on the calibration of the security systems/measures using linguistic scales with the aid of IDS software under a fuzzy environment. Pursuant to the evaluation results presented in Table 6.13, it can be seen that the highest amounts of 22.94 and 6.53% belief degrees of the High and Very High evaluation grades are associated with the access control measures in comparison to advanced cargo information process, screening measures, detection measures, personnel security and operational security.

The analysis has shown that personnel security measures has the lowest percentage in terms of belief degrees associated with the evaluation grades High and Very High. It is worth mentioning that comparison can also be made with respect to the Very Low and Low evaluation grades of the six security attributes. Based on the assessment, the final security risk level is obtained as 0.44 or 44%. This value represents the experts' assessment of the port's security and can be used to assist port security analysts to carry out security assessment of a port, and also initiate safety audit and review of key performance indicators of various port departments for enhanced operation. This approach may provide a viable approach where there is lack of objective or statistical data in such an assessment for the port under investigation.

6.4.6 Perform Sensitivity Analysis

Sensitivity analysis is performed to investigate how a change in the strength of evidence affects the results of the security estimates. Based on the axioms discussed in Sect. 6.3.6, the input data (i.e. belief degrees) of each attribute at the bottom level of the hierarchy are varied by 10%, 20% and 30% respectively and the corresponding risks' estimates are recorded. Accordingly, 57 experiments were conducted based on the axioms and by using the IDS software, the results obtained are presented in Table 6.14.

Based on the results in Table 6.14, a graph depicting the sensitivity of the analysis is drawn and presented in Fig. 6.5. The variations as a result of slight changes indicate that the model is sensitive to guards at the main security gate (rank 1), importer security filing (rank 2), watch-man at the quay/vessel interface (rank 3), perimeter intrusion detection device (rank 4), and biometrics (rank 5). The result of the analysis conclusively indicated that the least sensitive parameters of the model are monitoring device for warehousing (rank 19), monitoring device for quayside operations area (rank 18) and monitoring device for main gate (rank 17).

If the degrees of belief associated with the highest linguistic terms of all the lower level criteria are decreased by 0.15, the output data is assessed as 0.3178. By selection of 14 lower-level criteria (i.e. R24, R32, R34, R26, R21, R37, R39, R35, R27, R22, R33, R30, R31 and R38) from 19 and by decreasing the degrees of belief associated with the highest linguistic terms of these 14 lower level criteria by 0.15, the output data is evaluated as 0.3671. In view of the fact that 0.3671 is greater than 0.3178, the result is aligned with axiom 2.

Table 6.14 Experimental results for sensitivity analysis using the IDS software

Security risk attributes	Output data (10%)	Output data (20%)	Output data (30%)
Guards at main security gate	0.1988	0.3089	0.4142
Importer security filing	0.3325	0.4131	0.4935
Watch men at quay/vessel interface	0.3771	0.4426	0.5192
Perimeter intrusion detection device	0.3952	0.4572	0.5220
Biometrics	0.4054	0.4694	0.5290
Radiation detection device	0.4091	0.4722	0.5318
Gamma/X-ray detection measures	0.4177	0.4734	0.5328
Ocean vessel carrier filing	0.4200	0.4745	0.5339
Under-water and sonar cameras	0.4298	0.4798	0.5323
Perimeter fencing patrol	0.4347	0.4792	0.5310
Marshalling area patrol	0.4409	0.4870	0.5388
Water borne area patrol	0.4493	0.4881	0.5348
Cargo handling area patrol	0.4475	0.4896	0.5367
Access route patrol	0.4554	0.4907	0.5374
Monitoring device for ware house	0.4601	0.5016	0.5389
Monitoring device for stacking area	0.4721	0.5028	0.5379
Monitoring device for perimeter side	0.4782	0.5039	0.5369
Monitoring device for main gate	0.4829	0.5129	0.5409
Monitoring device for quayside operations area	0.4999	0.5209	0.5449

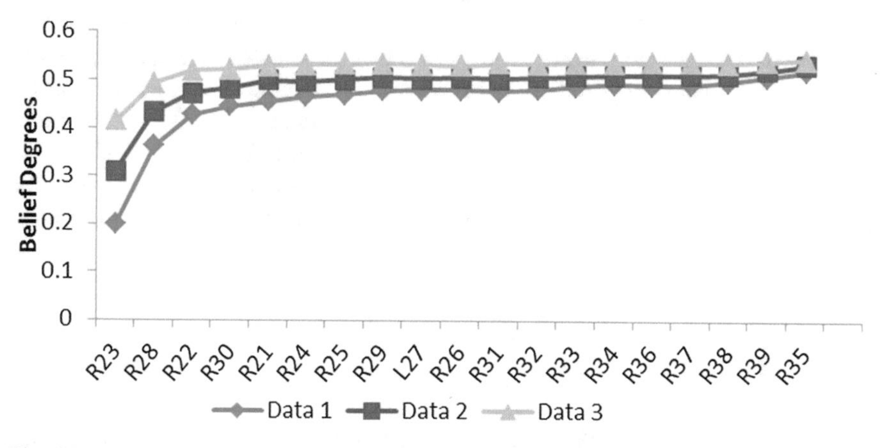

Fig. 6.5 Sensitivity analysis result conducted by varying the belief degrees

6.4.7 General Discussions

This security risk evaluation has shown the sensitivity of the model's parameters to slight changes in input data (see Fig. 6.5). The results of the experiments show that the performance of port security depends on many variables and a change in these variables will ultimately affect the security posture of ports operating in a dynamic environment. The model attempts to address maritime operations and their associated facilities and infrastructure as they attempt to collectively represent one of the single greatest unaddressed challenges to the security of nations and the global economy. The results of the analysis highlighted the significance of access control measures to seaport security given the dynamic nature of operations taking place within the system and also the importance of developing a comprehensive and effective regime of control over who and what will have access into the port environment. Additionally, another factor that influenced the performance effectiveness of seaport operations is the advanced cargo information process (R4). Based on the analysis, it can be shown that "advanced cargo information process" has the belief degree of 0.1963, which is associated with the High linguistic variable, and also significant in influencing seaport security. Based on industrial experience, the ISF and OVCF help the CBP to evaluate and identify high-risk cargo shipments going into a seaport in order to prevent smuggling and ensure cargo safety and security of infrastructure systems, it can be concluded that the model is sound and has revealed the influence magnitude of each parameter in a systematic fashion.

In the research conducted by Riahi [38], Talas and Menachof [45] and Christopher [10], access control measures and advanced cargo information process are mentioned as significant security measures reported in the literature due to a port's composition and its operational environment. The studies further revealed that ports that build their security foundation on these concepts are well equipped to control risks and ensure the safety of maritime cargo. Thus, the ranking orders in Fig. 6.5 are consistent with those discussed by Riahi [38], Talas and Menachof [45] and Christopher [10]. This evaluation can be used to develop schemes or measures that can enhance the performance of security systems/measures and reduce their vulnerability to security incidents.

6.5 Conclusion

This paper proposes a flexible approach to assess the performance of security systems/measures regarded as a multi-attribute decision making problem under a fuzzy environment. An analysis process that incorporates fuzzy logic, belief degree concept and evidential reasoning is proposed. In this approach, the performance effectiveness of security systems/measures are analysed using fuzzy set modelling. This provides the infrastructure analysts with flexibility in articulating judgements about the security risk parameters such as likelihood of threats, vulnerability of the system and consequent severity. The evaluation process of the security estimate is obtained using an evidential reasoning algorithm. Such an approach provides analysts with a robust tool to make full use of the information provided at the bottom level of the hierarchy to generate the overall security estimate for decision-making processes in a transparent manner.

It is beneficial that the proposed framework is based on existing and tested methods. Specifically, the model can be used during the design process of green seaport construction and green deep-water tank farms' construction within a port's system where PFSOs and CSOs can work hand-in-hand to facilitate a resilient security scheme that can best suit the company's need and purpose. The model's strengths have been seen from its potential to obtain security estimate from risk analysis associated with the security assessment requirement in ISPS Code and also form the basis for developing a structured generic security management methodology in a systematic way.

References

1. Adler, R.M., Fuller, J.: An integrated framework for assessing and mitigating risks to maritime critical infrastructure. In: 2007 IEEE Conference on Technologies for Homeland Security, pp. 252–257. IEEE (2007)
2. An, M., Huang, S., Baker, C.: Railway risk assessment-the fuzzy reasoning approach and fuzzy analytic hierarchy process approaches: a case study of shunting at waterloo depot. Proc. Inst. Mech. Eng. Part F: J. Rail Rapid Trans. **221**, 365–383 (2007)
3. Anoop, M., Rao, K.B., Gopalakrishnan, S.: Conversion of probabilistic information into fuzzy sets for engineering decision analysis. Comput. Struct. **84**, 141–155 (2006)
4. Bakir, N.O.: A Brief Analysis of Threats and Vulnerabilities in the Maritime Domain. Springer, Managing Critical Infrastructure Risks (2007)
5. Barnes, P., Oloruntoba, R.: Assurance of security in maritime supply chains: conceptual issues of vulnerability and crisis management. J. Int. Manag. **11**, 519–540 (2005)
6. Belton, V., Stewart, T.J.: Multiple Criteria Decision Analysis: An Integrated Approach. Kluwer, Norwell, MA (2002)
7. Bichou, K.: The ISPS code and the cost of port compliance: an initial logistics and supply chain framework for port security assessment and management. Marit. Econ. Logist. **6**, 322–348 (2004)
8. Blümel, E., Boevé, W., Recagno, V., Schilk, G.: Ship, port and supply chain security concepts interlinking maritime with hinterland transport chains. WMU J. Marit. Aff. **7**, 205–225 (2008)
9. Buckley, J.J.: Fuzzy hierarchical analysis. Fuzzy Sets Syst. **17**, 233–247 (1985)
10. Christopher, K.: Port Security Management. Auerbach Publications (2010)
11. Clarke, R.A., Lynch, P., Publico, P.B.: LNG Facilities in Urban Areas: A Security Risk Management Analysis for Attorney General Patrick Lynch. Rhode Island, Good Harbor Consulting (2005)
12. Dempster, A.P.: A generalization of bayesian inference. J. R. Stat. Soc. B (Methodological), 205–247 (1968)
13. DOHS: National Strategy for Maritime Security. Washington, DC (2005)
14. Emerson, S.D., Nadeau, J.: A coastal perspective on security. J. Hazard. Mater. **104**, 1–13 (2003)
15. Frittelli, J.: Port and maritime security: background and issues for congress. Port Marit. Secur. **11** (2008)

16. Greenberg, M.D., Chalk, P., Willis, H.H., Khilko, I., Ortiz, D.S.: Maritime Terrorism: Risk and Liability. Rand Corporation (2006)
17. Hecker, J.Z.: Port Security: Nation Faces Formidable Challenges in Making New Initiatives Successful. DTIC Document (2002)
18. Huang, C.L., Yoon, K.: Multiple Attribute Decision Making Methods and Applications: A State of Art Survey. Springer, New York (1981)
19. John, A.: Proactive risk management of maritime infrastructure systems: resilience engineering perspectives. Ph.D. thesis, Liverpool John Moores University-UK (2013)
20. Li, Y., Liao, X.: Decision support for risk analysis on dynamic alliance. Decis. Support Syst. **42**, 2043–2059 (2007)
21. Mansouri, M., Nilchiani, R., Mostashari, A.A.: Risk management-based decision analysis framework for resilience in maritime infrastructure and transportation systems. IEEE Syst. Conf. 35–41 (2009)
22. Mantaras, R.L.D.: Approximate Reasoning Models. Prentice Hall PTR (1990)
23. McGill, W.L., Ayyub, B.M., Kaminskiy, M.: Risk analysis for critical asset protection. Risk Anal. **27**, 1265–1281 (2007)
24. Mokhtari, K., Ren, J., Roberts, C., Wang, J.: Decision support framework for risk management on sea ports and terminals using fuzzy set theory and evidential reasoning approach. Expert Syst. Appl. **39**, 5087–5103 (2012)
25. Mostashari, A., Nilchiani, R., Omer, M., Andalibi, N., Heydari, B.: A cognitive process architecture framework for secure and resilient seaport operations. Mar. Technol. Soc. J. **45**, 120–127 (2011)
26. Moteff, J.: Risk Management and Critical Infrastructure Protection: Assessing, Integrating, and Managing Threats. Vulnerabilities and Consequences. Library of Congress Washington DC Congressional Research Service, DTIC Document (2005)
27. Nakamuara, K., Ogawa, Y., Kato, H., Shibasaki, R.: Impacts of maritime transportation risk on maritime traffic flows and regional 6 economies: the case study at the straits of Malacca and Singapore 7. In: Transportation Research Board 91st Annual Meeting (2012)
28. Ngai, E., Wat, F.: Fuzzy decision support system for risk analysis in e-commerce development. Decis. Support Syst. **40**, 235–255 (2005)
29. Omer, M., Mostashari, A., Nilchiani, R., Mansouri, M.: A framework for assessing resiliency of maritime transportation systems. Marit. Policy Manag. **39**, 685–703 (2012)
30. Orbeck, E.: Implementation of the ISPS Code in Norwegian Ports and Harbour Protection Through data Fusion Technologies. Springer Science Business Media B.V (2009)
31. Parfomak, P.W., Frittelli, J.: Maritime Security: Potential Terrorist Attacks and Protection Priorities. Congressional Research Service, DTIC Document, Washington DC (2007)
32. Pate, A., Taylor, B., Kubu, B.: Protecting America's Ports: Promising Practices. Police Executive Research Forum, 221075 (2007)
33. Percival, B.: Indonesia and the United States: Shared Interests in Maritime Security. United States-Indonesia Society (2005)
34. Pfingsten, T.: Bayesian active learning for sensitivity analysis. In: Proceedings of the 17th European Conference on Machine Learning, pp. 353–364. Springer, Berlin, Germany (2006)
35. Pillay, A., Wang, J.: Technology and Safety of Marine Systems. Elsevier Science (2003)
36. Price, W.: Reducing the risk of terror events at seaports 1. Rev. Policy Res. **21**, 329–349 (2004)
37. Raymond, C.Z.: Maritime terrorism in Southeast Asia: a risk assessment. Terrorism Polit. Violence **18**, 239–257 (2006)
38. Riahi, R.: Enabling security and risk-based operations of container liner supply chains under high uncertainties. Ph.D. thesis, Liverpool John Moores University (2010)
39. Riahi, R., Bonsall, S., Jenkinson, I., Wang, J.: A seafarer's reliability assessment incorporating subjective judgements. Proc. Inst. Mech. Eng. Part M J. Eng. Marit. Environ. **226**, 313–334 (2012)
40. Roy, B., Vanderpooten, D.D.: The European school of MCDA: emergence, basic features, and current works. Eur. J. Oper. Res. **99**(1), 26–27 (1997)
41. Saaty, T.L.: Analytic Hierarchy Process. McGraw-Hill, New York (1980)

42. Sadiq, R., Kleiner, Y., Rajani, B.: Water quality failures in distribution networks-risk analysis using fuzzy logic and evidential reasoning. Risk Anal. **27**, 1381–1394 (2007)
43. Shafer, G.: A Mathematical Theory of Evidence. Princeton University Press Princeton (1976)
44. Steward, T.J.: A critical survey on the status of multiple criteria decision making theory and practice. OMEGA Int. J. Manag. Sci. **20**, 569–586 (1992)
45. Talas, R., Menachof, D.A.: The efficient trade-off between security and cost for sea ports: a conceptual model. Int. J. Risk Assess. Manag. **13**, 46–59 (2009)
46. Tang, M.T., Tzeng, G.H., Wang, S.W.: A hierarchy fuzzy MCDM method for studying electronic marketing strategies in the information. service industry. J. Intell. Inf. Manage. **8**(1), 1–22 (2000)
47. UNISYS: White Paper on Port Security Roadmap. Unisys Corporation, Blue Bell, PA, USA (2011)
48. Voss, M.D., Whipple, J.M., Closs, D.J.: The role of strategic security: internal and external security measures with security performance implications. Trans. J. 5–23 (2009)
49. Yager, R.R.: On the determination of strength of belief for decision support under uncertainty–Part II: fusing strengths of belief. Fuzzy sets Syst. **142**, 129–142 (2004)
50. Yang, J.B., Singh, M.G.: An evidential reasoning approach for multiple attribute decision making with uncertainty. IEEE Trans. Syst. Man Cybern. **24**(1), 1–18 (1994)
51. Yang, J.B., Xu, D.L.: On the evidential reasoning algorithm for multiple attribute decision analysis under uncertainty. IEEE Trans. Syst. Man Cybern. Part A Syst. Humans **32**, 289–304 (2002)
52. Yang, Z., Wang, J., Bonsall, S., Fang, Q.: Use of fuzzy evidential reasoning in maritime security assessment. Risk Anal. **29**, 95–120 (2009)
53. Yang, Z., Wang, J., Li, K.: Maritime safety analysis in retrospect. Marit. Policy Manag. **40**(3), 261–277 (2013)

Chapter 7
Maritime Simulation Using Open Source Tools: Ship Transits in Bosporus

Murat M. Gunal

Abstract Maritime transportation is one of the most significant components of the world's economy and therefore safer, efficient, and sustainable transportation systems are essential. In the design of these systems, Operational Research/Management Science methods can help decision makers at operational and strategic levels. Simulation is one of the methods in the toolbox with its proven characteristics including scalability, flexibility, and accountability. Although simulation has been used for analyzing maritime transportation systems before, and there are examples in the literature, simulation model building process and models developed are not explicitly published. To fill this gap, and guide model builders in maritime transportation domain, this chapter presents a step by step development of a model which simulates maritime traffic in Bosporus, a narrow and busy strait in Istanbul, Turkey. The model utilizes two open source libraries in Java; OpenMap, a geographical information system, and SimKit, a discrete event simulation library. The model demonstrates the relationships between sea traffic rules, number of pilots, and waiting times. This chapter presents a simple and an extended version of the simulation model. The simple version includes one type of ship arrival, one pilot, and one radar. The extended version is a scaled-up version where these entities are multiplied; two types of ship arrivals, five pilots, and three radars. The models are fully customizable and can be tailored for various purposes. For illustration, the extended version is used to analyse the effects of change in number of pilots and mean of interarrival times to waiting times of ships.

Keywords Simulation · Istanbul straits · Queuing analysis

7.1 Introduction

The volume of maritime traffic is increasing every year and its management for international community is becoming more problematic than it was in the past. In

M.M. Gunal (✉)
Industrial Engineering Department, Turkish Naval Academy,
Tuzla, Istanbul, Turkey
e-mail: m_gunal@hotmail.com

C. Konstantopoulos and G. Pantziou (eds.), *Modeling, Computing and Data Handling Methodologies for Maritime Transportation*, Intelligent Systems Reference Library 131, DOI 10.1007/978-3-319-61801-2_7

the last decades, new problems and concerns emerged in Maritime Transportation (MT), such as maritime piracy and environmental effects. Transportation companies are looking for safer and cleaner routes for their ships. On a wider perspective, all stakeholders are looking for the ways of protection against piracy for safer routes, as well as of reducing carbon emission gases for cleaner environment.

Operational Research/Management Science (OR/MS) methods and tools can help provide insights to these problems. Decision makers in MT can tackle these problems, and others, using optimisation, simulation, and analytics techniques which OR/MS has been providing for many years. In fact, the birth of modern OR comes from the sea. During World War II, OR techniques were used to intercept submarines in open sea. Since its birth, new methods have been developed and these are applied in wider sectors.

One of the popular techniques in OR/MS is simulation. Simulation is a method in which real systems are mimicked on computer. Since doing this is safer and cheaper than trying in real, it is adopted and applied in many domains. Healthcare, production, service systems, and transportation benefit from simulation.

The purpose of this chapter is to demonstrate how to build a basic simulation model of maritime operations for decision making. This chapter is written as a tutorial rather than a text to explain the concepts in simulation modelling. The readers are assumed that they are already aware of basic simulation modelling concepts and capable of programming in Java language.

In this tutorial, two open source libraries are used to build a model of maritime transportation;

- OpenMap (http://www.openmap-java.org/)
- SimKit (https://eos.nps.edu/simkit/)

The simulation model utilizes OpenMap for creating a front-end (e.g. user interface and graphical features) and a backbone for geographical data manipulation. Since it is an open source software, the user can extend its default appearance by adding new features such as text, icons, and images. Second, SimKit package is utilized to simulate entities on map. Simulated entities (e.g. Ships) interact with each other and with the map. The model is based on conceptualisation of movement and detection in SimKit [4].

The model in this exercise is a fictional case on Bosporus, the narrow strait in Istanbul, Turkey. The Bosporus is one of the busiest waterways in the world which connects Black Sea and Marmara Sea. It is 17 nautical miles in length and passes through Istanbul city. It includes sharp turns and due to the ships carrying hazardous material there is high risk of accident. Turkish Strait Vessel Traffic Services (TSVTS) system is established in late 1990s and modernized in 2006. Its purpose is to enhance the maritime traffic and environmental safety against risk and dangers which may be resulted from maritime traffic, within specified service area in Turkish Straits including the Bosporus [7]. The TSVTS includes 16 unmanned traffic surveillance stations with radars, and there are pilotage services provided to ships which transits in south and north bounds.

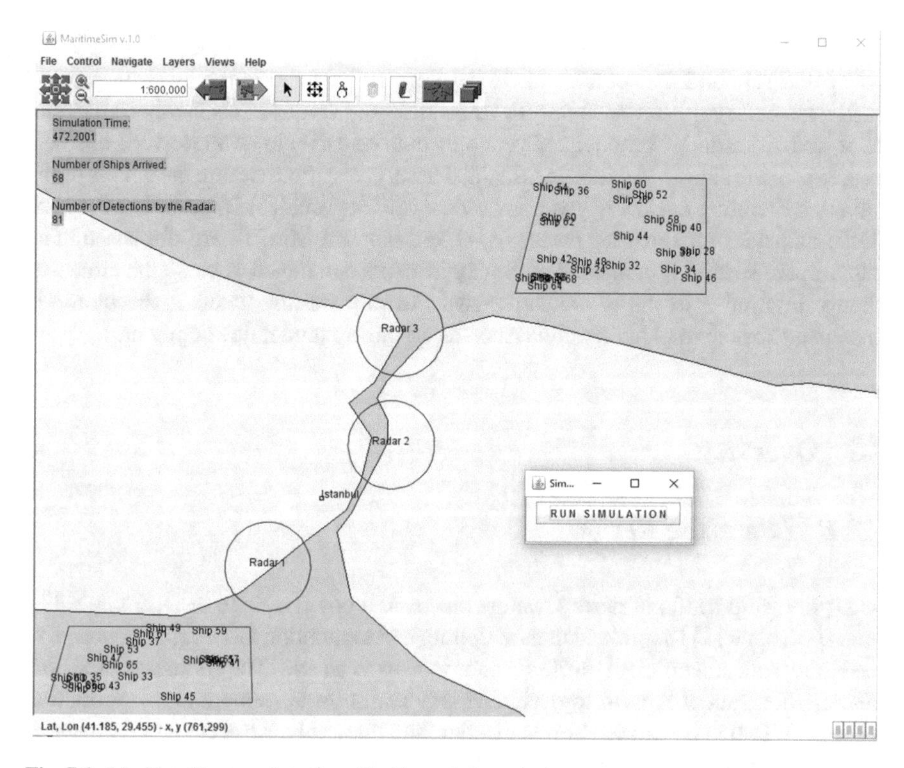

Fig. 7.1 MaritimeSim user interface (the Extended version)

MaritimeSim is a simulation model which simulates transit of ships through Bosporus. In this tutorial, first, a simple version of the model is built step by step and at the end, an extended version is presented with the resultant user interface as shown in Fig. 7.1. In the simple version, ships arrive Yenikapi region randomly, the red area on the bottom of the map, in which interarrival times follow exponential distribution. In the extended version, ship arrivals also occur in the North. The model assumes that there are 10 pilots which accompany ships and therefore the transit is possible if one pilot is available for a ship. In other words, the Bosporus is assumed to be a multi-server queuing system. Note that in the simple version, we have only one pilot working. When ships reach the end of the strait, either in the South (Marmara Sea) or in the North (Blacksea), the pilot spends approximately one hour to return to his base in the North or in the South. However, the simple version simulates only one-way (North bound) traffic. To add an extra feature to simulation, there are radars at three locations which works independently and can detect a ship passing by. The radar is included to mimic the TSVTS system component and its role here is just to count the number of ships passed through.

Full source code (Simple and extended versions), and executables of the model is given at the link www.simulationmodel.com/MaritimeSim.

In the following sections, there are specifics of MaritimeSim including how it is set up and run, and the source code is developed. For making the learning process easier, the code explanation is only done for the simple version (one-way traffic, one pilot, and one radar). The extended version would be easier to be understood once the concepts in the simple version, as explained here, is clarified. Before the details of the model, a literature review is given, overview of Geographical Information Systems (GIS) and the two software packages, OpenMap and SimKit, are discussed. The penultimate section is dedicated to the experiments conducted to study the effects of change in number of pilots and ship arrivals to ship waiting times. A discussion is also given for the implementation issues faced during model development.

7.2 Overview

7.2.1 Literature Review

Analysing ship traffic in narrow waters has been a popular topic in recent decades. Ince and Topuz [11] expressed that modelling and simulation has a significant role in designing safe and efficient traffic systems in narrow waters. They demonstrated how a simulation model is used to design Vessel Traffic Management and Information System (VTMIS) in Turkish Straits. Earlier than this study, Kose et al. [12] analysed effects of new oil pipelines in Black Sea, and potential increase in number of ships passing Istanbul strait. Their findings affected the maritime administration to run one-way traffic flow in Bosporus. Almaz et al. [1] also simulated Bosporus in greater detail, considering natural factors such as visibility and water current, resources such as pilotage and tug boat services, and management rules such as overtaking rules. Their study echoes previous ones; the increasing risk of ship trans passing Bosporus. Basar [2] simulated ship traffic in Çanakkale Strait to examine also the effects of increase in volume on waiting times of ships and risk of transit. His study focused on a specific area in the strait.

Although in these studies Commercial-Off-The-Shelf simulation software is used, Huang et al. [10] pointed out that there is no adequate simulation tool to study maritime traffic on ports and narrow waters. They developed a systematic approach, MTSS, that enables developing simulation models of complex traffic on busy ports. They showed its use in a busy port and conducted experiments under emergency situations such as closure of waterways. Their tool helped port authority professionals discover the impacts of such emergencies. MTSS had also been used for analysing ship traffic in Tokyo Bay, Straits of Singapore and Shanghai [9]. Singapore Straits has also been analysed by [17] using a different approach.

Other than simulation, different methods are also used for analysis. Mavrakis and Kontinakis [14] presented a queuing model to analyse Bosporus strait ship traffic. Yip [19] proposed a traffic flow model for maritime domain using differential equations. The mathematical model considers water current and ship domain concepts which can analyse two-dimensional marine traffic on a water way. Ucan and Nas [15] took a similar view of Bosporus and studied pilot capacity. They showed that number of pilots must increase due to the diversity of ships and increasing traffic in Bosporus.

With regards to risk, Geijerstam and Scensson [8] is a detailed study which investigates ship collision risk in general, and ship accidents in offshore oil installations in specific. The factors involve humans, technical equipment, and decision making at several levels of organisational structure.

7.2.2 *Geographical Information Systems (GIS)*

A GIS is software to deal with the earth's geography. There are many GIS packages and tools in the market. ARCView, Google Maps, GeoKIT, NASA World Wind, Digital Nautical Chart, Falcon View, and OpenMap are some of the known GIS software. Among these software packages some of them require licence but others are freely available and open source, such as OpenMap. OpenMap is a free and open source software written in Java programming language. This is the main purpose why this GIS is chosen in this modelling exercise, first, it is free and open source, and second it is written in Java and therefore compatible with the simulation software package SimKit.

GIS software manipulates geographical digital data. What makes a GIS software valuable is the data supplied to it. It is like a satellite navigation on a car, it is useless without the map data. If, for example, you are interested in vehicle routing on roads, then the road data, which is generally represented by vectors, must be given to the software. Three types of digital map data exist;

- Vector maps
- Elevation maps
- Raster maps

A vector map consists of vectors, or lines, which their start and end points are known. Think about shore lines or political boundaries, or rivers, or roads in a town. You can draw these shapes with lines, either straight or curved on a paper. Curved lines are more difficult to represent on computer since they need more parameters than a straight line needs. For a straight line, or say for multiple straight and connected lines, you need to keep starting point and ending point of each line. By a "point" we meant a point on earth which means a pair of latitude and longitude. On a 2-dimensional computer screen it is easy to covert world coordinate system (Lat-Lon pair) to screen coordinate system (X, Y pair). On the other hand, curves can be approximated by a number of small line segments.

There are different formats of elevation data. The common one is the Digital Terrain Elevation Data (DTED). DTED simply keeps grid based elevation data, that is the earth is divided into squares and for each square there is an elevation value. Dimensions of these squares are matter of resolutions and in standards there are three levels; in DTED level 0 (DTED0), a square's one side is 900 meters, in DTED level 1 (DTED1) it is 90 meters, and in DTED level 2 it is 30 meters.

Raster maps are paper based maps which are scanned and converted to digital format. It is convenient to see a real map on screen but in digital format it is not of much use, since the data is actually a scanned image.

As mentioned earlier a GIS software manipulates the data supplied to it. It can handle different data at once. The data generally displayed as "layers", which you can turn on and off. For example, you can show a vector map together with DTED map. In this chapter, a simulation model is also represented as a layer in OpenMap. Therefore, you can observe simulation with the map data.

7.2.3 OpenMap and Simkit

OpenMap package is an open source GIS package written in Java. It includes an API and is customizable to create GIS based Java applications. Although it offers GIS functions necessary for developing applications, programmers can add new features and link other packages. OpenMap is fully open source and ideal for geospatial analysis.

Simkit package is a Discrete Event Simulation (DES) code library also written in Java. It is developed by [5]. Although it is a generic DES simulation package, it performs better in modelling defence systems such as naval operations. This package is taught at graduate level simulation course in the U.S. Naval Postgraduate School (NPS). There are notes and examples on the website on how a Simkit model can be built. Furthermore, Simkit is one of the open source simulation APIs [6] available.

There is no direct link between these two pieces of software. However, as this tutorial shows, it is possible to link them two and use together. This link is first established in a master's thesis at NPS [13] and showed that they can work well together. SimKit and a different GIS (GeoKit) is used in another study [18] where military deployment operations are analysed. Buss and Ahner [3] is one of the applications of SimKit with GIS capabilities. They built a DES model, using OpenMap and SimKit, to find optimal allocation of firing and sensors of military deployment. In their study, they discuss advantages and disadvantages of level of detail in models and conclude that a low-level approach is suitable for fast and flexible model building.

Distinction of this modelling exercise is that it uses ready-made classes of SimKit and OpenMap. Therefore, it allows fast and minimum effort modelling. However, the modeller can still extend the work here, for example by adding rules of interaction to vessels. This can be achieved by using "Event Graphs", the founding concept of SimKit [5]. An Event Graph is the representation of simulation events on a graph diagram and is ideal for implementing Event Scheduling (ES) algorithm. It has events as circles to represent nodes and transitions as arrows to represent edges. When an event occurs in simulation, it triggers other events to occur after a delay in the future. An EG is drawn for this case study.

Since OpenMap and Simkit are written in Java, we will develop our model in Java. We are going to use Eclipse (http://eclipse.org) compiler environment. Attachment source files are developed and tested at Eclipse IDE for Java Developers, version Mars.2.Release (4.5.2).

7.3 MaritimeSim Simulation Model

Maritime Simulation Model (MaritimeSim) is built to simulate ship movements in Bosporus. It implements the EG in Fig. 7.2 and includes 5 events. MaritimeSim also creates events implicitly from SimKit's Sensor and Mover classes. Note that the events here are executed by the simulation engine and are unique to each ship entity and therefore it is an abstraction. In MaritimeSim, Java code is written for each event.

"Run" event is the initial event which starts the simulation. As soon as it is executed at time 0, a Ping event is generated. The role of Ping is to animate simulation entities on map, and it basically draws all simulation entities on their current location. Moving entities change position in time relative to their speed and route. Ping event is a recursive event which schedules itself every inter ping time (IPT). IPT determines the screen updating rate. Run event also triggers "Ship Arrival" event which causes new ships to be added to simulation. It is also recursive and is executed every inter arrival time (IAT). IAT is an ideal approach to generate ship arrivals randomly such as from Exponential distribution with an average IAT. In the model, we assumed that one ship arrives every 15 min. The role of Ship arrival is to generate new ships and include them to simulation. Arriving ships' initial positions are also determined randomly in Yenikapi region marked as red lines on the bottom of Fig. 7.1. Upon arrival, ships wait for a pilot to accompany the transit. The "$\sim$" sign between Ship Arrival and Start Transit events is the condition to execute Start Transit event. This event causes the ship to move and follow predefined Bosporus route. This route is already entered in the code and include Lat and Lon pairs of waypoints along the route. When the ship starts, it triggers "End Transit" and "Start Move" events. We can schedule the "End Transit" event as soon as ship starts moving since we assume that ship moves linearly along the route. The condition between End Transit and Start Transit is for checking the queue of ships waiting for the pilot, since a Transit can only occur if a pilot is available.

Connection between MaritimeSim's events and SimKit is the link between Start Transit and "Start Move". Start Move is implicitly generated by signalling the Mover object in simulation. Role of Start Move is to make the calculation necessary to find End Move time. Start and End Move events trigger each other since the ship moves along waypoints linearly and therefore its position and time on that position can be calculated. SimKit includes linear movement calculations. We know exact positions of initially generated entities such as the Radar. As soon as a ship starts moving, it calculates when and where it will enter the Radar's range. "Enter Range" event is triggered once a ship enters the detection range of Radar. Simultaneously it triggers "Detection" event to occur. There is only a counter inside the Detection event. Whilst

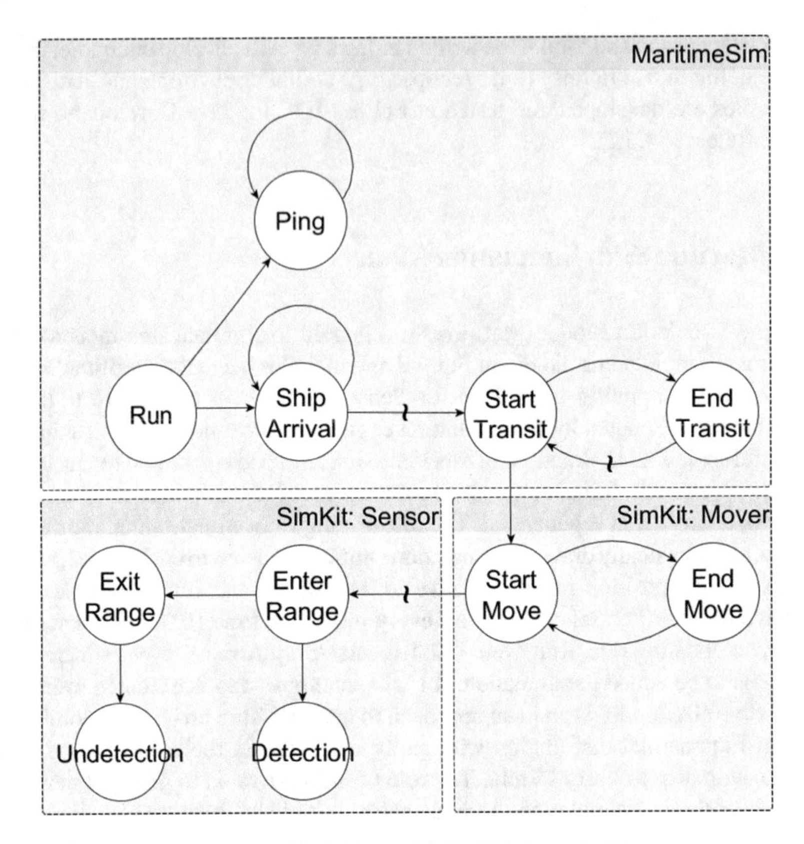

Fig. 7.2 Event graph of maritime simulation model (the Simple version)

the ship is moving inside the Radar's range circle, its Exit time can be calculated and be scheduled as "Exit Range" which also means "Undetection".

The EG in Fig. 7.2 is conceptual representation of the system we modelled. Its implementation is done in MaritimeSim and will be explained in following sections. MaritimeSim includes 5 files which work together (Table 7.1). First two files are text files to run and to setup the application. Last three files are Java source code files in MaritimeSim model.

7.3.1 "openmap.properties" File

"openmap.properties" file is the input file for OpenMap GIS package. The file starts with initialization parameters (Code-Box 1). Lines 10 to 12 are indicating center point of map's initial position, and its initial scale. Once the application is started, the user can change these as needed. Line 15 is the projection used on the map. Width

Table 7.1 Files in the Maritime Model

File name	Explanation
openmap.properties	Properties file for OpenMap which includes setup values
MaritimeSim.bat	Windows batch (script) file to run the application
SimulationLayer.java	OpenMap layer to display simulated entities on the map. This file is the interface between OpenMap and the SimKit model
SimulationExtraLayer.java	OpenMap layer to display extra graphical objects on the map. This works as a layer on the map
SimulationModel.java	Simulation model to simulate entities on the map. This works as a thread on the map

and Height of the map window is specified on lines 18 and 19, and the title is set on line 22.

Code-Box 1: openmap.properties input file—part 1.

```
8     # Latitude and longitude in decimal degrees
9     # Istanbul (Bosphorus) Strait
10    openmap.Latitude=41.046f
11    openmap.Longitude= 29.023f
12    openmap.Scale=600000.0f
13
14    # Projection type to start the map with.  Try "cadrg",
    "orthographic",
15    openmap.Projection=cadrg
16
17    # Width and Height of map, in pixels
18    openmap.Width=1000
19    openmap.Height=700
20
21    # Change this for a different title in the main window.
22    openmap.Title=MaritimeSim v.1.0
```

This file is a long file which include every detail necessary to setup the GIS application. However for MaritimeSim, very few is required. Lines 640 to 644 (Code-Box 2), at the end of the file, specifies the class names of the two layers which will be included in MaritimeSim application. ".classs" property is "MartimeSim." text and this shows the Java package name developed for the model. This package will be explained in Java files section.

Code-Box 2: openmap.properties input file—part 2.

```
640    SimulationLayer.class=MaritimeSim.SimulationLayer
641    SimulationLayer.prettyName=Simulation
642
643    SimulationExtra.class=MaritimeSim.SimulationExtraLayer

644    SimulationExtra.prettyName=Simulation Extras
```

After indicating the location of class files, you must then tell OpenMap that you want to show these layers. To do this you need to add "SimulationLayer" and "SimulationExtra" words in the "openmap.layers" string on Line 220 and 230. These layers are shown in the same order as they are written here. This is important since the mouse control is given to the top layer. Note that as class and prettyName properties are written for SimulationLayer, each of these layers have also property lines in the input file. A layer may also have extra properties that are entered in the input file. Simulation Layer has only these default two properties. If you want to parameterize this layer, then you need to write "getProperties" method in your layer class.

OpenMap is a GIS software and it can only work with the GIS data, such as vector maps, Digital Terrain Elevation Data (DTED) maps, raster maps. These maps are actually files on your computer's file system and it is your task to tell OpenMap where these files are located. MaritimeSim is set to display vector maps (ShapePolitical layer) and the data is included with the application. Line 299 indicates that the map data is under "data/shape/cntry02" folder.

7.3.2 Running the Application and the "MaritimeSim.bat" File

To run the application, rather than running from Eclipse, "MaritimeSim" batch file is recommended (Code-Box 3). This file includes DOS commands to set up necessary libraries, such as Java and OpenMap. The user must edit line 10 according to the location where the application is installed.

When the user interface appears, to start the simulation, you need to open SimulationLayer's control panel. To do this, click on the left most mini button on lower right corner of the screen. It will open "Simulation" window with "Run Simulation" button which will start the simulation.

Code-Box 3: MaritimeSim.bat batch file.

```
1    @echo off
2
3    rem Java Virtual Machine
4    set JAVABIN=java.exe
5
6    rem MaritmeSim folder
7    rem ************************************************
8    rem Only change MaritmeSim_HOME folder!
9
10   set MaritimeSim_HOME=C:\MaritimeSim
11
12   rem Below this line, all paths are relative to _HOME
   path
13   rem No need to change anything.
14   rem ************************************************
15
16   set OPENMAP_HOME=%MaritimeSim_HOME%\lib\openmap.jar
17
18   set CLASSPATH=%OPENMAP_HOME%;
19   set CLASSPATH=    %CLASSPATH%;
                       %MaritimeSim_HOME%\inputs\;
                       %MaritimeSim_HOME%\bin\;
                       %MaritimeSim_HOME%\lib\;
                       %MaritimeSim_HOME%\lib\simkit.jar;
20
21   rem Run MaritimeSim
22   %JAVABIN%      -mx256m      -Dopenmap.configDir=%Mari-
   timeSim_HOME% -                          Ddebug.showpro-
   gress com.bbn.openmap.app.OpenMap >    %Mari-
   timeSim_HOME%\out.txt
23   rem  to write the output to a file: > %Mari-

   timeSim_HOME%\out.txt
```

7.3.3 *"SimulationLayer.java" File*

This file creates a GIS layer for OpenMap which communicates with simulation model. Java imports are required libraries. The other two parts, OpenMap and Simkit, are the other libraries that this class needs. This layer implements several interfaces. For example, SimEventListener is used to implement the "processSimEvent" method, which will be discussed later.

Code-Box 4: SimulationLayer class imports and class definition.

```
1     package MaritimeSim;
2
3     import java.awt.Color;
4     import java.awt.event.ActionEvent;

      ......

14    import com.bbn.openmap.Layer;
15    import com.bbn.openmap.event.*;

      ......

27    import simkit.SimEvent;
28    import simkit.SimEventListener;

      ......

33    public class SimulationLayer extends Layer implements
34  SimEventListener,
                              MapMouseListener {
```

The constructor's role is to instantiate the simulation model and to display the radar. The "simModel" variable is the link between the model and the map layer. The rest of the constructor is the code required for showing Radar's graphics on map. In constructor, radarText and radarCircle graphic objects are created on the location specified in model file.

Code-Box 5: SimulationLayer class constructor.

```
53    public SimulationLayer(){
54
55     simModel = new SimulationModel(0.1, 100, false);
56
57     Point2D p1=simModel.getLocationRadar();
58
59     radarText=new OMText((float)p1.getX(),(float)p1.getY(),"Radar",1);
60     radarCircle=new OMCircle((float) p1.getX(),(float)p1.getY(),
                          simModel.getRangeRadar());
61     radarCircle.setLinePaint(Color.BLUE);
62    }
```

Most critical method of the SimulationLayer class is "processSimEvent". This method is overridden from SimEventListener. It listens to events generated by SimKit and process these events accordingly. Each simulation event is fetched with its name, for example Line 73 handles "Ping" event. The ping event as shown in Fig. 7.2 is triggered as soon as the model starts, and repeatedly schedules itself. When a Ping event is heard all ship's circles lat-lon locations are updated. Line 75 is the loop to iterate on Simulation model's movers dictionary. This dictionary keeps ships' data generated (arrived) in the model. The iterator variable nM is a temporary variable. "shipsCircle" and "shipsText" are also dictionaries as the same size as "movers" dictionary. On Line 78, the ship's circle is picked from "shipsCircle" dictionary and its Lat and Lon is updated from the model's mover equivalent. The same thing is done for the text of ships on lines 81 and 82.

Code-Box 6: SimulationLayer class processSimEvent method.

```
64    @Override
65    public void processSimEvent(SimEvent e) {
66    fireStatusUpdate(LayerStatusEvent.START_WORKING);

67    /*
68    if (e.getEventName().equals("Run")) {
69
70    }*/
71
72    if (e.getEventName().equals("Ping")) {
73
74        for (Enumeration<String> mM = simModel.movers.keys();
75                                      mM.hasMoreElements();) {
76            String shipName =(String) mM.nextElement();
77            shipsCircle.get(shipName).setLatLon(
78            (float) simModel.getLocationMover(shipName).getX(),
79            (float) simModel.getLocationMover(shipName).getY());

80            shipsText.get(shipName).setLat((float)
81            simModel.getLocationMover(shipName).getX());
82            shipsText.get(shipName).setLon((float)
83            simModel.getLocationMover(shipName).getY());
84        }
85    }

86    if (e.getEventName().equals("ShipArrival")) {
87
88        nShipsArrived++;
89
90        Object[] par=e.getParameters();
91
```

```
 92            String shipName="Ship "+par[0].toString();
 93
 94            Point2D p1= simModel.getLocationMover(shipName);
 95
 96            text1=new  OMText((float)  p1.getX(),  (float)p1.getY(),ship-
        Name,1);
 97            circle1=new OMCircle((float) p1.getX(),(float)p1.getY(),1,1);
 98            circle1.setLinePaint(Color.BLUE);
 99
100            shipsCircle.put(shipName, circle1);
101            shipsText.put(shipName, text1);
102        }
103
104        if (e.getEventName().equals("Detection")) {
105
106            nDetection++;
107
108        }
109
110        if(proj != null){
111
112            displayObjectsOnMap();
113
114        }
115        repaint();
116        fireStatusUpdate(LayerStatusEvent.FINISH_WORKING);
117       }
```

"ShipArrival" event's role is to add circle and text objects to this layer. On Line 86, when ShipArrival event is heard, the ship's location is got from the model on line 94. Since the simulation model already generated a mover object, its location is known in model and the rest is just to create shadow circle and text objects. After these objects are created, on line 112, they are displayed on map.

"displayObjectsOnMap" method displays all map graphic objects on map by using OpenMap's "generate" method. This method visualizes the graphic object on the current projection of the map.

"projectionChanged" method is overridden for OpenMap's Layer class. This method is essentially executed when the map is panned or zoomed. The "paint" method works with this method. Its role is to render graphic objects.

Code-Box 7: SimulationLayer class display, projection change and paint methods.

```
119     private void displayObjectsOnMap(){
120      for (Enumeration<String> mM = shipsCircle.keys(); mM.hasMoreEl-
      ements();)      {
121         String key = (String) mM.nextElement();
122         shipsCircle.get(key).generate((Projection)proj);
123            shipsText.get(key).generate((Projection)proj);
124      }
125      radarText.generate((Projection)proj);
126      radarCircle.generate((Projection)proj);
127     }
128
129     //@Override
130     public void projectionChanged(ProjectionEvent e) {
131      proj = e.getProjection();
132      System.out.println("projection Changed");
133
134      displayObjectsOnMap();
135      repaint();
136     }
137
138     public void paint(java.awt.Graphics g) {
139
140         for (Enumeration<String> mM = shipsCircle.keys(); mM.hasMo-
      reElements();)  {
141         String key = (String) mM.nextElement();
142            shipsCircle.get(key).render(g);
143            shipsText.get(key).render(g);
144        }
145
146        radarCircle.render(g);
147        radarText.render(g);
148        fireStatusUpdate(LayerStatusEvent.FINISH_WORKING);
149     }
150
151     public double getSimTime(){
152      return simModel.getSimTime();
153     }
```

Finally the last part of our simulation layer is the control panel of this layer. There could be a menu item or a panel for controlling the simulation. "getGUI" method is written to create a simple user interface and to start the simulation. On line 158, when this method is called, simModel variable starts to send simulation events to this layer, or in other words this layer listens to simulation model. The same listener linkage is done for mover and radar objects in lines between 160 and 167. On line

169, the “Run Simulation” button’s action is managed. When this button is pressed, simulation model’s “startPinging” method is called which activates the thread.

Code-Box 8: SimulationLayer class getGUI method.

```
155        //A GUI for the layer
156        public java.awt.Component getGUI() {
157           JPanel returnPanel = new JPanel();
158           simModel.addSimEventListener(this);
159
160           for (Enumeration<String> mM = simModel.movers.keys();
                                         mM.hasMoreElements();) {
161
162           String key = (String) mM.nextElement();
163
164              simModel.movers.get(key).addSimEventListener(this);
165          }
166
167           simModel.radar.addSimEventListener(this);
168
169           runButton.addActionListener(new ActionListener() {
170              public void actionPerformed(ActionEvent e) {
171                    simModel.startPinging();
172              }
173          });
174
175           returnPanel.add(runButton);
176
177           return returnPanel;
178         }
```

7.3.4 “SimulationModel.java” File

This file is the simulation model class file which runs as a thread. It includes necessary Java and SimKit classes and extends “SimEntityBase” which is the base class for any SimKit model. The “Runnable” class is a Java class and its role is to make this simulation model run as a thread. A Thread is related to the operating system, Microsoft Windows in our case, to be able to run a programme in parallel to other programmes. This makes the simulation run parallel to map application and therefore does not lock the map during simulation run.

Code-Box 9: SimulationModel class imports and class definition.

```
1     package MaritimeSim;
2
3     import java.awt.geom.Point2D;
4     import java.util.*;
5
6     import simkit.*;
7     import simkit.random.*;
8     import simkit.smdx.*;
9
10    import com.bbn.openmap.event.DistanceMouseMode;
      ……
17    public class SimulationModel extends SimEntityBase implements
   Runnable {
```

The constructor takes three arguments (Code-Box 10). "deltaT" is the "inter ping time" which means simulation time step for screen updates. In every "Ping" event (in "doPing" method), all simulation object graphics are updated to their current location. Second argument in the parameter string is "millisPerSimtime" which determines sleep time of simulation thread. This causes the model to suspend for a small amount of time and to continue running afterward. The last parameter is a Boolean flag which sets the animation on or off. On line 47, we set the number of pilots to one in this simulation model. Lines between 49 to 52, we instantiate a random variable generator. "RandomVariateFactory" on line 52 is a method in SimKit for creating generators. We create Exponential distribution generator with an average interarrival time of 15 min. Lines between 54 and 58 are related to referee mechanism in SimKit. A SimKit entity communicates with other entities via "Mediators". A mediator is set between sensors and movers. By doing so, for example, a mover sends messages to the referee (ref variable) and this referee broadcasts this message to those who are concerned.

"routeLatLon" is a floating point variable array and the numbers in it are the coordinates of the Bosporus route. This route is applied to all ships transiting Bosporus. The for loop in lines 74 to 77 fills the "bosphRoute" array list with these numbers.

There is one radar in the simulation model and it includes a mover object. A sensor (the radar) must be with a mover object in SimKit and the "dummyRadarMover" in line 82 takes this role. Since it is stationary, its speed is set to 0. In line 85, we set this mover to start automatically once the model is reset. Additionally, in line 86, we register radar to the referee. The same logic is applied to all other movers (ships) in the loop in line 89.

SimKit events are shown in Code-Box 11. There are as many event handlers as the events shown in Fig. 7.2 MaritimeSim part. The first one is "Run" event's "doRun" method. For each event in an event graph, a "do[Event Name]" method must be written in SimKit. "doRun"'s function is to start the simulation thread by calling "startPinging" method. This is explained in Code-Box 12. Second event is the “Ping” event and its handler is “doPing”. Inside this method, if the animation is on, or the isPinging method returns true, a “waitDelay(“Ping”, deltaT)” command in line 118 is executed. “waitDelay” command is a SimKit command to schedule a new event in event list in simulation. This is the implementation of self loop in Fig. 7.2 “Ping” event. Line 118 causes new ping events to be scheduled every deltaT time unit and when this method is executed, it also causes simulation to sleep for “deltaT * millisPerSimtime” real time units, in line 120. Screen updating occurs in Simulation Layer class, since the graphic objects are belonging to that class.

Code-Box 10: SimulationModel class constructor.

```
41  public SimulationModel(double deltaT, double millisPerSimtime, boolean
42                  pinging) {
43
44      setMillisPerSimtime(millisPerSimtime);
45      setDeltaT(deltaT);
46      setPinging(pinging);
47
48      nPilot = 1;                             // We have 1 pilot on hand
49
50      String distribution = "Exponential"; // Initialise the distribu
51                                 tion for     inter-arrival times
52      Object[] param = new Object[1];        // of ships
53      param[0] = new Double(15);             // Average interarrival times
                                   is in   minutes
54      arrivalTimeGenerator = RandomVariateFactory.getInstance(distribu-
55  tion, param);

56      SensorTargetMediatorFactory.addMediator(     //  Initialize   the
57  referee
58              CookieCutterSensor.class,
59               UniformLinearMover.class,
60               CookieCutterMediator.class);
61      ref = new SensorTargetReferee();
62
63        /* The Bosphorus Route */
64      float[] routeLatLon= new float[] {
65              40.967587f, 28.94353f,
66              41.023632f, 29.012878f,
67              41.05202f, 29.038883f,
68              41.090595f, 29.063927f,
69              41.13063f, 29.076448f,
70              41.152466f, 29.058147f,
71              41.185947f, 29.08897f,
72              41.247818f, 29.142908f
```

```
73      };
74
75        bosphRoute = new ArrayList<WayPoint>(); // Create way points from
                                                 // Lat and Lon pairs
76      for (int i = 0; i < routeLatLon.length/2; ++i) {// and add them to
77          WayPoint  w=new  WayPoint(new  Point2D.Float(routeLatLon[2*i],
78                                  routeLatLon[2*i+1]), 0.1);
79                 bosphRoute.add(w);
80          }
81
82        /* The Radar
           * It must have a stationary (velocity 0) mover object.*/
83        Point2D radarLatLon= new Point2D.Float(41.22097f, 29.103422f);
84        UniformLinearMover dummyRadarMover = new UniformLinearMover("Ra
                     dar", radarLatLon, 0.0);
85      radar=new CookieCutterSensor(nmToDeg(1,2.0f), dummyRadarMover);
86        PatrolMoverManager manager = new PatrolMoverManager(dummyRadar
87         Mover,    bosphRoute);
88        manager.setStartOnReset(true);
89        ref.register(radar);

90      // Get ready for moving the mover objects
91        for (Enumeration<String> mM = moverManagers.keys();
92                                          mM.hasMoreElements();) {
93         String key = (String) mM.nextElement();
94             moverManagers.get(key).setStartOnReset(true);
95          }
96    }
```

"doShipArrival" is the longest event handler. The parameter ID value is the ship ID and it is passed between Ship Arrival event to keep track of the ID number of the last ship arrived. Next ship's arrival is scheduled in line 162 and its parameter, for the next arriving ship, is incremented by one. If the current ship's ID is 1, the next ship's ID will be 2 and its arrival will occur after amount of time the "arrivalTimeGenerator" generates in line 162. "doShipArrival" starts with generating a latitude value between 40.8754 and 40.9452, lines 128 and 129. Likewise, a random uniform value is generated for a longitude value. These are the left-right, and upper-lower boundaries of Yenikapi anchoring region. By doing this, we create a random arrival location (line 135). When the ship starts to move from this initial location, it will follow Bosporus route, and will proceed until the final location. Line 142 creates a final location in the middle of the Black Sea. After preparing for the new mover, it is created in line 146 by adding to "movers" dictionary. Movers dictionary's key is a string containing ship's name, and its value is a "UniformLinearMover". This is a SimKit class and a basic implementation of movement. A mover starts from an initial location, and has a speed. However its movement is managed by a "Mover Manager". For each mover, there must be a mover manager. Mover manager is responsible for movement discipline, such as patrolling, random walk, following a path. All movers in this simulation model follows the Bosporus route. Since the bospRoute is a dynamic array, it adds up all arriving ship's final destinations. Therefore (in line 158) the final destination is removed every time it is added to the route. Line 160 is the queue implementation. A ship's ID is added to a queue so that the ship in turn will be picked from that queue. As explained, line 162 is for scheduling next ship's arrival. The condition in line 164 is for the start of ship's movement. A ship can only start its transit if there is a pilot available and there is a ship waiting in queue. Once this condition is satisfied, we schedule in line 165 a "ShipTransit" event straight away, for the ship we have just pushed to queue. Also, start of transit means that a pilot is engaged with this ship's transit so that we decrement the number of available pilots by one, in line 166.

Code-Box 11: SimulationModel class SimKit Event handler methods.

```
104     /*****************************************************
105      *
106      *     SIMKIT Events
107      *
108      *****************************************************/
109
110     public void doRun() {
111     if (this.isPinging()) {
112         startPinging();
113     }
114     }
115
116     public synchronized void doPing() {
117     if (isPinging()) {
118         waitDelay("Ping", deltaT);
119          try {
120             Thread.sleep((long) (deltaT * millisPerSimtime));
121          } catch (InterruptedException e) {}
122     }
123     }
124
125     public void doShipArrival(int ID){
126
127     UniformVariate uLat= new UniformVariate();
128     uLat.setMaximum(40.8754);
129     uLat.setMinimum(40.9452);
130
131     UniformVariate uLon= new UniformVariate();
132     uLon.setMaximum(28.9154);
133     uLon.setMinimum(28.6845);
134
135     Point2D arrivalLocation=new Point2D.Double(   uLat.generate(),

136     uLon.generate());
```

```
137
138     UniformVariate uLonFinal= new UniformVariate();
139     uLonFinal.setMaximum(32.5);
140     uLonFinal.setMinimum(28.5);
141     float fnum=new Float(uLonFinal.generate());
142
143     WayPoint   wFinal=new   WayPoint(new   Point2D.Float(42.0f,   fnum
144  ),0.166);
145       bosphRoute.add(wFinal);
146
147
148     movers.put("Ship "+ID, new UniformLinearMover("Ship "+ID,
149                             arrivalLocation,
150                             0.166));
151
152     ref.register(movers.get("Ship "+ID));
153
154     moverManagers.put("Ship "+ID, new PathMoverManager(
155                             movers.get("Ship "+ID),
156                             bosphRoute));
157
158     moverManagers.get("Ship "+ID).setStartOnReset(true);
159
160     bosphRoute.remove(wFinal);
161
162     shipQueue.add(ID);
163
164     waitDelay("ShipArrival",arrivalTimeGenerator.generate(), ID+1);
165     if (nPilot > 0 && !shipQueue.isEmpty()) {
166         waitDelay("StartTransit", 0.0, shipQueue.remove());
167         nPilot--;
168     }
169     }
170
171     public void doStartTransit(int ID){
172     moverManagers.get("Ship "+ID).start();
173     waitDelay("EndTransit",  96.0, ID);
174     }
175
176     public void doEndTransit(int ID){
177     nPilot++;
178
179     if (nPilot > 0 && !shipQueue.isEmpty()) {
180         waitDelay("StartTransit", 0.0, shipQueue.remove());
181         nPilot--;
182     }
183     }
```

Fourth event is "StartTransit" event which is implemented in line 170. Since it is certain that the movement will start, the mover manager with ship ID is started in line 171. Once a mover is started, it is certain that it will end its transit after 96 units of time. This is the amount of time a ship transits Bosporus. It could be a random time but for simplicity a deterministic time is chosen. The "EndTransit" event handler's task is to release the pilot by incrementing number of available pilots by one, and to start next ship's transit. Note that this is clear in EG in Fig. 7.2, the link between EndTransit and StartTransit events. The If statement in line 178 is checking if there is available pilot and there is ship waiting to transit. If this condition is satisfied, we schedule a new transit event and decrease the available number of pilots by one.

The other major methods in Simulation Model class is shown in Code-Box 12. "startPinging" method essentially start the main thread which causes "run" method to be executed. "run" method is different than "doRun" method because it is called by the thread but not by SimKit. In line 199, we schedule first ship arrival (ID 1) at time 0 and first ping event at time 0.0001. This little gap between first ship arrival and first ping is to put these two events in order. The ping event will be executed slightly after ship arrival so that a deadlock will not happen. The loop in line 202 is to start all mover managers. Commands in lines 209 to 214 are all related to starting a SimKit model. "setVerbose" method causes simulation to print event list in every step of simulation execution. This is useful to keep track of events execution. As explained in "Running the Application" section, the output will be written to command line in Windows, or to a text file, depending on the last line on "MaritimeSim.bat" batch file. "stopAtTime" method is to set the run length of simulation. "1440" is the minutes in a day. Finally, line 213 actually starts the simulation.

Code-Box 12: SimulationModel class thread methods.

```
190     // Method to start the thread
191     public void startPinging() {
192      new Thread(this).start();
193     }
194
195     // This method is called when the thread starts
196     public void run() {
197      setPinging(true);
198
199      waitDelay("ShipArrival", 0.0, 1); // Schedule the arrival of the
    first ship at time 0,
200      waitDelay("Ping", 0.0001);    // and with the ID 1. Schedule the
    first Ping event
201
```

```
202          for (Enumeration<String> mM = moverManagers.keys();
                                   mM.hasMoreElements();;) {
203              String key = (String) mM.nextElement();
204              moverManagers.get(key).start();
205        }
206
207            if (!Schedule.isRunning()) {
208
209              System.out.println("Simulation starts");
210            //Schedule.setSingleStep(false);
211              Schedule.setVerbose(true);
212              Schedule.stopAtTime(1440.0);
213            Schedule.startSimulation();
214            System.out.println("Simulation ends");
215        }
216       }
```

7.3.5 *"SimulationExtraLayer.java" File*

This file is an extra file to display some objects on map. It is separated from Simulation Layer to differentiate indirect needs in simulation, such as drawing a border for anchoring area, and outputs of simulation. It is a layer just like Simulation Layer, and its visibility can be set on or off.

Code-Box 13: SimulationExtraLayer imports and definition.

```
1      package MaritimeSim;
2
3      import java.awt.*;
4      import java.text.*;
5
6      import com.bbn.openmap.event.*;
7      import com.bbn.openmap.layer.LabelLayer;
8      import com.bbn.openmap.omGraphics.*;
       ......
13     public class SimulationExtraLayer extends LabelLayer {
```

The constructor of this layer includes a polyline graphic object. This is used to mark Yenikapi anchoring area. The "paint" method renders this area and makes some simulation outputs displayed on the upper left corner of the map. These are simulation time, number of ships arrived, and number of detections made by the radar.

Code-Box 14: SimulationExtraLayer constructor and paint method.

```
27      public SimulationExtraLayer(){
28      super();
29      omList=new OMGraphicList();
30
31      OMPoly  p2  =  new  OMPoly(waitingArea, OMGraphic. DECIMAL_DEGREES,

32      OMGraphic.LINETYPE_STRAIGHT);
33          p2.setLinePaint(Color.red);
34          omList.add(p2);
35      }
36
37      public void paint(Graphics g) {
38          DecimalFormat df = new DecimalFormat("###.########");
39
40          String data = df.format(simLayer.getSimTime());
41
42          labelText = "Simulation Time:\n"+data;
43          labelText+="\n\nNumber of Ships Arrived:\n"+ simLayer.nShipsAr
              rived;
44          labelText+="\n\nNumber of Detections by the Radar:\n"+
                                                        simLayer.nDetection;
45
46          if(omList.size() > 0){
47              omList.render(g);
48          }
49
50          super.paint(g);
        }
```

7.4 Experiments and Discussion

The extended version of the model is used for experimental analysis. Although a range of factors and levels are available to be included in the analysis, the experiments comprise two factors; full-time equivalent number of pilots, and arrival volumes, and five levels in each factor. The first factor is one of the capacity constraints regarding waiting times. This factor is a manageable factor since the transit management can hire more pilots to ease the waiting times of ships in Bosporus. Figure 7.3 shows the waiting times of ships for five levels of number of pilots. In this figure, waiting times are shown as the averages and confidence intervals at 95% level (bars) for 10 simulation replications. Since the variation is not high, 10 replications are enough for obtaining meaningful results. Inter arrival times on both directions are assumed to be 25 min on average.

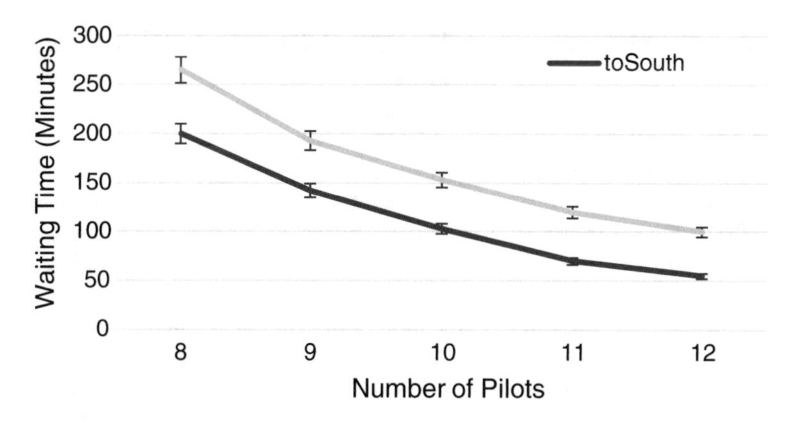

Fig. 7.3 Waiting times of ships by number of pilots

Ships heading to North, from Marmara to Black Sea, have higher waiting times. This is due to the strong currents on this direction and longer transit times. In the model, transit times, or the service times in queuing terms, are different in North and South directions to reflect the effect of water current. As expected, when the number of servers increase waiting times decrease, and vice versa. However, the exponential shape suggests that we will need more pilots to diminish the waiting time.

The second factor, mean interarrival times of ships, is included to explore potential effects of the Istanbul Channel. Opening a man-made channel on the far west of Bosporus is an idea developed in the last decade and aims at easing the ship traffic load on Bosporus. Historic values show that on both directions inter arrival time is 25 min on average. This value is increased by 10 min-steps and its effects on waiting times are shown in Fig. 7.4. Note that by increasing this value, we are decreasing

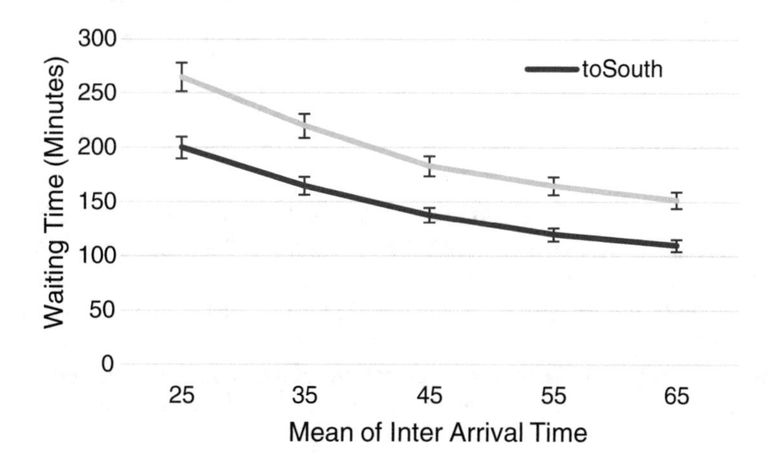

Fig. 7.4 Waiting times of ships by mean of inter arrival times

expected number of ships generated. The Istanbul Channel project is not definite yet, and its effect on inter arrival times are not known yet, but it is certain that ships' waiting times are likely to decrease in Bosporus, as the graph suggests. We assumed that there are 10 pilots working for both directions.

These two sets of experiments show the use of extended model. To do the experiments, new variables are added in the code to collect necessary statistics such as waiting times. On a ship arrival, we record the arrival time and on the exit, we find the difference and record this value. Changing number of pilots and inter arrival times are easy to do in the code.

The extended version is an amplification of the simple model. For example, ship arrival, start transit, end transit events in Fig. 7.2 are doubled in the extended version, since two-way traffic is simulated. For the ship arrival, "ShipArrivalNorthBound" and "ShipArrivalSouthBound" events generate new arrivals on appropriate locations and times. Likewise, "StartTransitNorthBound", "StartTransitSouthBound", "EndTransitNorthBound", and "EndTransitSouthBound" events are added in the code. The modeller must add necessary start and end events for any other type of movements. With regards to radar events, there is no implementation in the "SimulationModel.java" file since the events are generated automatically by the Simkit's Sensor and Mover objects.

MaritimeSim model is an example of developing a simulation model by programming. Note that there are several software available on the market, Commercials Off the Shelf (COTS), which have user interfaces and ready-made objects for fast model building. Using an API, such as SimKit and OpenMap, require programming expertise however give full flexibility to a modeller. Using a COTS or API to simulate something is a difficult decision. If the problem on hand include very special requirements, such as handling GIS and multi interacting entity types, COTS simulation software may not suffice. Varol and Gunal [16] is an example of such case in which the problem required handling GIS and interacting five entity (agent) types. MartimeSim model has similar requirements and therefore model building is done by programming.

7.5 Conclusion

Simulation is an effective method to evaluate a real system's performance on computer. Although there are many simulation software in the market, modellers still need programming to create customized simulation based decision support tools. The objective of this chapter is to teach, and guide to, how a maritime simulation model can be programmed in Java, using two open source packages. This tutorial includes source codes with detailed explanations of the model's simple version. The extended model is a scaled up version of this basic model and given as an online supplement.

In the tutorial, a model for a fictional case in Istanbul Strait is built. Ships arrive to anchoring area before transiting Bosporus and wait until a pilot is available. Ship transit the strait North bound (the basic version), and North and South bounds (the extended version) on a predefined route. At the end of the route, there is a radar which detects the passing ships in the basic version. The model animates ships on map and the number of arriving and transited ships are counted and displayed. A conceptual model of this system is created using Event Graph paradigm. EG is an ideal representation for DES.

In most of the simulation studies published in the academic literature, simulation models are not openly revealed. This chapter particularly contributes to this end, since every detail is explained and full source code is given. By doing so, we expect emergence of new studies which use simulation method explained here, to create value in maritime industry.

References

1. Almaz, A.O., Or, I., Ozbas, B.: Investigation of transit maritime traffic in the strait of Istanbul through simulation modeling and scenario analysis. Int. J. Simul. Syst. Sci. Technol. (7) No. 7 (2006)
2. Basar, E.: Investigation into marine traffic and a risky area in the Turkish straits system: canakkale strait. Transport **25**(1), 5–10 (2010)
3. Buss, A.H., Ahner, D.K.: Dynamic allocation of fires and sensors (Dafs): a low-resolution simulation for rapid modeling. In: Perrone, L.F., Wieland, F.P., Liu, J., Lawson, B.G., Nicol, D.M., Fujimoto, R.M. (eds.) Proceedings of the 2006 Winter Simulation Conference (2006)
4. Buss, A.H., Sanchez, P.: Simple movement and detection in discrete event simulation. In: Kuhl, M.E., Steiger, N.M., Armstrong, F.B., Joines, J.A. (eds.) Proceedings of the 2005 Winter Simulation Conference, pp. 992–1000 (2005)
5. Buss, A.H.: Basic Event Graph Modeling. Technical Notes, Simulation News Europe, 1–6 April 2001
6. Dagkakis, G., Heavey, C.: A review of open source discrete event simulation software for operations research. J. Simul. Adv. Online Publ. 19 June 2015. doi:10.1057/jos.2015.9
7. Emniyeti, K., Mudurlugu, G.: Vessel Traffic and Pilotage Services. www.kiyiemniyeti.gov.tr/default.aspx?pid=23 (2016). Accessed June 2016
8. Geijerstam, K.A.F., Scensson, H.: Ship Collision Risk-An identification and evaluation of important factors in collisions with offshore installations. Departmen of Fire and Safety Engineering and Systems Safety. Lund University, Sweden. Report 5275 (2008). ISSN: 1402-3504
9. Hasegawa, K., Yamazaki, M.: Qualitative and quantitative analysis of congested marine traffic environment—an application using marine traffic simulation system. Int. J. Mar. Navig. Saf. Sea Trans. **7**(2) (2013)
10. Huang, S.Y., Hsu, W.J., Fang, H., Song, T.: MTSS—a maritime traffic simulation system and scenario studies for a major hub port. ACM Trans. Model. Comput. Simul. (TOMACS) **27**(1), Article No. 3 (2016)
11. Ince, A.N., Topuz, E.: Modelling and simulation for safe and efficient navigation in narrow waterways. J. Navig. **57**(1), 53–71 (2004)
12. Kose, E., Basar, E., Demirci, E., Guneroglu, A., Erkebay, S.: Simulation of marine traffic in Istanbul Straits. Simul. Model. Pract. Theory **11**(2003), 597–608 (2003)
13. Mack, P.: THORN: a study in designing a usable interface for a geo-referenced discrete event simulation. MSc Thesis, Naval Postgraduate School, USA (2000)
14. Mavrakis, D., Kontinakis, N.: A queueing model of maritime traffic in Bosporus Straits. Simul. Model. Pract. Theory **16**(2008), 315–328 (2008)

15. Ucan, E., Nas, S.: Analysing Istanbul strait maritime pilot capacity by simulation technique. J. Navig. **69**(4), 815–827 (2016)
16. Varol, A.E., Gunal, M.M.: Simulating prevention operations at sea against maritime piracy. J. Oper. Res. Soc. **66**(12), 2037–2049 (2015)
17. Xiaobo, Q., Meng, Q.: Development and applications of a simulation model for vessels in the Singapore Straits. Expert Syst. Appl. Int. J. **39**(9), 8430–8438 (2012)
18. Yildirim, U.Z., Tansel, B.C., Sabuncuoglu, I.: A multi-modal discrete-event simulation model for military deployment. Simul. Model. Pract. Theory, **17**(4), 597–611 (2009). doi:10.1016/j.simpat.2008.09.016, ISSN: 1569-190X
19. Yip, T.L.: A marine traffic flow model. Int. J. Mar. Navig. Saf. Sea Trans. **7**(1) (2013)

Printed in the United States
By Bookmasters